SpringerBriefs in Earth System Sciences

Series Editors

Gerrit Lohmann, Universität Bremen, Bremen, Germany

Lawrence A. Mysak, Department of Atmospheric and Oceanic Science, McGill University, Montreal, QC, Canada

Justus Notholt, Institute of Environmental Physics, University of Bremen, Bremen, Germany

Jorge Rabassa, Labaratorio de Geomorfología y Cuaternar, CADIC-CONICET, Ushuaia, Tierra del Fuego, Argentina

Vikram Unnithan, Department of Earth and Space Sciences, Jacobs University Bremen, Bremen, Germany

SpringerBriefs in Earth System Sciences present concise summaries of cutting-edge research and practical applications. The series focuses on interdisciplinary research linking the lithosphere, atmosphere, biosphere, cryosphere, and hydrosphere building the system earth. It publishes peer-reviewed monographs under the editorial supervision of an international advisory board with the aim to publish 8 to 12 weeks after acceptance. Featuring compact volumes of 50 to 125 pages (approx. 20,000—70,000 words), the series covers a range of content from professional to academic such as:

- A timely reports of state-of-the art analytical techniques
- bridges between new research results
- snapshots of hot and/or emerging topics
- literature reviews
- in-depth case studies

Briefs are published as part of Springer's eBook collection, with millions of users worldwide. In addition, Briefs are available for individual print and electronic purchase. Briefs are characterized by fast, global electronic dissemination, standard publishing contracts, easy-to-use manuscript preparation and formatting guidelines, and expedited production schedules.

Both solicited and unsolicited manuscripts are considered for publication in this series.

More information about this series at https://link.springer.com/bookseries/10032

Johanna Fink · Elisa Heim · Norbert Klitzsch

State of the Art in Deep Geothermal Energy in Europe

With Focus on Direct Heating

 Springer

Johanna Fink
Institute for Applied Geophysics
and Geothermal Energy
RWTH Aachen University
Aachen, Germany

Elisa Heim
Institute for Applied Geophysics
and Geothermal Energy
RWTH Aachen University
Aachen, Germany

Norbert Klitzsch
Institute for Applied Geophysics
and Geothermal Energy
RWTH Aachen University
Aachen, Germany

ISSN 2191-589X ISSN 2191-5903 (electronic)
SpringerBriefs in Earth System Sciences
ISBN 978-3-030-96869-4 ISBN 978-3-030-96870-0 (eBook)
https://doi.org/10.1007/978-3-030-96870-0

This Springer imprint is published by the registered company Springer Nature Switzerland AG
The registered company address is: Gewerbestrasse 11, 6330 Cham, Switzerland

Preface

As the Earth continuously supplies heat, geothermal energy is a renewable energy source. Since nearly 50% of Europe's energy demand is in the heating and cooling sector (Bertani et al. 2018), it is expected that geothermal energy will play an important role in the transition to a decarbonized energy system.

Currently, geothermal energy in the European Union has an installed capacity of about 24.3 GW_{th} for heating and cooling (Kujbus et al. 2020). This is only a very small amount compared to the total geothermal potential of Europe (ETIPDG 2018; Garabetian 2019). In 2018, Europe had 280 geothermal district heating systems, spread over 24 countries with an installed capacity of 4.8 GW_{th}. However, geothermal energy is currently harvested mainly from high-enthalpy resources, i.e. from resources located in regions with favorable geothermal conditions. Because these areas are geographically limited, using geothermal energy in less favorable regions with low-to-medium enthalpy resources, is essential for unleashing the full potential of geothermal energy, as these resources account for the majority of Central Europe's total geothermal potential (Richter 2017).

This book project arose from a request made by E.ON for a literature review about the state of the art in deep geothermal energy. It relates to an EU call for the demonstration of innovative technologies to reduce greenhouse gas emissions. In this call, E.ON applied for funding for the application of deep geothermal for district heating. We believe E.ON's request reflects the growing interest in deep geothermal energy in general, and among energy companies and communities in particular. In this book, we introduce the opportunities and limitations of geothermal energy usage, particularly in geothermally less favorable regions. To this end, we review the state of the art in deep geothermal energy with focus on direct heating. The book thus provides an overview of technologies used to obtain heat from the deep underground and discusses main technical and non-technical risks associated with deep geothermal projects.

Addressing readers without a geoscientific background, we first lay the foundations of geothermal energy by explaining the heat distribution in the Earth as well as the governing geological and geophysical processes in Chap. 1. Moreover, we introduce the classification of geothermal systems to set the frame for the geothermal potential across Europe, which we discuss subsequently in Chap. 2. After a general overview, we explain in detail the geothermal potential of three example regions (Sweden, Poland, and Hungary) with different geological conditions. Chapter 3 gives an overview of the technologies applied for harnessing heat from the subsurface. We introduce the geothermal project development process and focus on exploration, assessment, development, and operation of a geothermal reservoir. For non-favorable subsurface conditions, reservoir stimulation is required for producing heat. Therefore, we introduce methods of developing these so-called Enhanced Geothermal Systems (EGS) and give an overview of existing large-scale EGS projects in Europe. We highlight innovative technologies and assess their potential to make heat usable. Associated technical and non-technical risks and barriers are outlined in Chap. 4. Finally, we summarize the findings of our literature study and draw conclusions regarding the technological state of the art for deep geothermal energy and its potential for direct heat use in Europe.

This work was initiated and funded by E.ON. We would like to thank our colleagues Bérénice Vallier, Matthis van Wickeren, and Tobias Ganther for their support with research and writing. Additionally, we thank Thorsten Gleu for the design of some of the figures and several publishers and authors for their permission to use and reproduce images.

Aachen, Germany   Johanna Fink
October 2021   Elisa Heim
                                                                      Norbert Klitzsch

References

Bertani R, Buesig H, Buske S, Dini A, Hjelstuen M, Luchini M, Manzella A, Nybo R, Rabbel W, Serniotti L, Science TD, Team T (2018) The First Results of the DESCRAMBLE Project. In: PROCEEDINGS, 43rd Workshop on Geothermal Reseroir Engineering. Stanford University, Stanford, California, p 16, URL https://pangea.stanford.edu/ERE/pdf/IGAstandard/SGW/2018/Bertani.pdf. Accessed 25 Oct 2021

ETIP-DG ETaIPoDG (2018) Strategic Research and Innovation Agenda. Tech. rep., European Technology and Innovation Platform on Deep Geothermal (ETIPDG). URL http://www.etip-dg.eu/front/wp-content/uploads/AB_AC_ETIP-DG_SRA_v3.3_web.pdf. Accessed 25 Oct 2021

Garabetian T (2019) Report on Competitiveness of the geothermal industry. Tech. rep., European Technology and Innovation Platform on Deep Geothermal (ETIP-DG). URL http://www.etip-dg.eu/front/wp-content/uploads/D4.6-Report-on-Competitiveness.pdf. Accessed 25 Oct 2021

Kujbus A, van Gelder G, Urchueguia JF, Pockelé L, Guglielmetti L, Bloemendal M, Blum P, Pasquali R, Bonduà S (2020) Strategic Research Innovation Agenda for Geothermal Technologies. Tech. rep., Geothermal Panel of European Technology Platform on Renewable Heating and Cooling. URL www.rhc-platform.org. Accessed 21 October 2021

Richter M (2017) Summary of New Drilling Technologies. Tech. rep., International Energy Agency - Geothermal (IEA Geothermal). URL http://iea-gia.org/wp-content/uploads/2014/10/IEA-Geothermal-Drilling-Technologies.pdf. Accessed 25 Oct 2021

Contents

Chapter 1
Introduction to Geothermal Systems

Abstract The distribution of heat in the Earth's crust is determined by various geophysical and geological processes. As heat constantly flows from the Earth's interior to the surface, temperature increases with depth. The rate of temperature change with depth is called geothermal gradient. A high geothermal gradient results in a high geothermal potential, thus making a region attractive for the use of geothermal energy. The geothermal gradient of a region depends on the predominant type of heat transfer in the subsurface, and the thermal and hydraulic rock properties. Based on these geological controls, a distinction is made between hydrothermal and petrothermal systems, both of which require different strategies for the technical development of the geothermal reservoir.

Keywords Geothermal reservoir types · Geothermal play types · Geothermal gradient · Heat flow · Heat transport · Thermal rock properties

1.1 Introduction

Geothermal energy is heat stored below the Earth's surface. On the surface, this energy manifests itself, for example, in the form of volcanoes and thermal springs. Thermal water has been used for heating and bathing for thousands of years. Today, even though geothermal energy can be used to generate electricity, it is primarily used for geothermal heating and cooling. While near-surface geothermal energy (up to 400 m depth) can be used only for heating in combination with a heat pump, deep geothermal energy can be used directly at temperatures above 20 °C. Fluid temperatures of above 120 °C qualify for power generation. In between these temperature levels, the application of geothermal energy ranges from heating and cooling of individual buildings over district heating to agricultural and industrial processes. In addition, some innovative applications include geo-cooling, melting snow or ice, and seawater desalination.

The type of use and the potential of geothermal plants strongly depend on the exact location on Earth. Because heat is not uniformly distributed in the Earth' crust, some sites are more favorable for geothermal energy extraction than others. Before we

© The Author(s), under exclusive license to Springer Nature Switzerland AG 2022
J. Fink et al., *State of the Art in Deep Geothermal Energy in Europe*,
SpringerBriefs in Earth System Sciences,
https://doi.org/10.1007/978-3-030-96870-0_1

delve into the geothermal potential of Europe in the next chapter, we first introduce the geological and geophysical processes that lead to the distribution of heat in the Earth's crust. Next, we explain the classification of geothermal systems (play types) from a geological perspective and the three different geothermal reservoir types.

1.2   Earth's Heat

The global surface temperature is 15 °C. This temperature, however, increases with depth. The Earth's inner core records temperatures of up to 6000 °C. The internal heat originates from two main sources. About 30% stems from the residual heat of the Earth's interior due to accretion when Earth was formed, and its slow but consistent cooling. The remaining around 70% mostly result from the decay of radioactive isotopes in the mantle and the crust (Stober and Bucher 2020). Unfortunately, current drilling technologies are unable to reach the high temperatures in the inner Earth. For example, to date, the deepest borehole has a depth of about 12 km, which covers only 0.2% of the Earth's radius (6370 km).

Even though we can only scratch the surface, temperature increases with depth. The hot interior of the Earth causes a heat flow towards the surface, thus creating a **temperature gradient** (geothermal gradient). The mean gradient in the crust is about $30 \,^{\circ}\text{C km}^{-1}$. However, Earth's crust varies in regard to thickness, especially due to the movement of tectonic plates relative to each other. The movement results in the building of mountain ranges and the thickening of crust, or in continental breakup including volcanic activity and the thinning of the crust. Depending on the thickness of the crust and the location on a tectonic plate, there are both positive and negative geothermal gradient anomalies. For example, in extensional settings such as at mid-ocean ridges, which have a thin crust, the geothermal gradient can reach more than $100 \,^{\circ}\text{C km}^{-1}$.

Similar to the spatial variation of the geothermal gradient, the rate per unit area that vertically flows from the Earth's interior to the surface can be analyzed. It is called **heat flux**, heat flow density, or geothermal flow (Pasquale et al. 2017). The mean heat flux on continental crust amounts $67 \,\text{mW m}^{-2}$ (Pollack et al. 1993). In Europe, the average is $65 \,\text{mW m}^{-2}$. Heat flow is, analogous to the geothermal gradient, generally highest close to plate boundaries, for example, at mid-ocean ridges like Iceland ($1500 \,\text{mW m}^{-2}$) and elevated around young mountain ranges ($300 \,\text{mW m}^{-2}$) or sedimentary basins ($100 \,\text{mW m}^{-2}$) (e.g. Toth 2016). Heat flux is a function of the geothermal gradient and the thermal conductivity of the rocks in the subsurface, i.e. their ability to transport or conduct heat, and thus strongly dependent on geologic controls.

These spatial variations of heat flux are caused by heat transport in the subsurface. Two main types of heat transfer mechanisms are distinguished:

1. **Conductive heat transport** is the dominant type of heat transfer in the crust. It depends on the temperature gradient and the thermal conductivity of the rocks and thus on their composition (see Sect. 1.3 for more details).
2. **Convective heat transport** takes place by movement of hot fluids, for example, water, gas or molten rock (magma). Examples would be the rise of magma through the crust and the emplacement of a granitic body or the circulation of fluids from deep fault zones to the surface resulting in thermal springs. The permeability of a rock mass, i.e. its ability to transmit fluids, is the controlling factor for convective heat transport.

From an economic point of view, areas with a positive anomaly of the geothermal gradient and high heat flux are particularly interesting and favorable for geothermal use, because higher temperatures are reached at shallower depths. This makes the accessibility and development of the geothermal resource easier and cheaper.

1.3 Rock Types and Their Properties

Geologically, three main rock types are distinguished based on their genesis:

- **Sedimentary rocks** are formed when particles from water or air settle or when minerals precipitate from a liquid. They are deposited in layers, such as sandstone or limestone, and are mostly unconsolidated after deposition. As more sedimentary layers are stacked on top of each other, the sediments beneath are buried. Burial results in compaction of the sediment grains, and minerals in between gradually cement them together forming a consolidated sedimentary rock.
- **Igneous rocks**, such as granite, are formed when molten rock cools and becomes solid, e.g. magma that rises upwards and emplaces in the crust as a granitic body.
- **Metamorphic rocks**, such as gneiss or marble, are formed when existing rocks are exposed to heat, pressure or mineral-rich fluids, which results in a change of their original properties, e.g. through partial melting or recrystallization.

All rocks are composed of various minerals in crystalline or amorphous appearance. Unlike sedimentary rocks, igneous and metamorphic rocks are mostly crystalline, which leads to a strong difference in their hydraulic and mechanical properties. The following paragraphs give an overview of rock properties and provide qualitative estimates of those properties for crystalline and non-crystalline (sedimentary) rocks. For more detailed information on rock properties, it is referred to Schön (2015).

Hydraulic properties comprise porosity and permeability of a rock mass. They play a decisive role for the heat transfer process that dominates in the geothermal system. Moreover, they are crucial for the achievable flow rate, which decides whether a geothermal reservoir needs to be enhanced by stimulation techniques. **Porosity** is the ratio of the pore volume to the total volume of a rock. If the pore network is interconnected and permits fluid to flow through the pore space, the rock additionally has an intrinsic or **intergranular permeability**. In sedimentary rocks, porosity and permeability are usually high after their deposition and decrease during burial. Sedimentary rocks are generally considered as reservoir rocks for water, oil, and gas. In addition, **fracture permeability** allows rocks that do not have a natural pore system, such as crystalline rocks, to be permeable as well. Fractures are linked to discontinuities and often result from tectonic stresses, which also define the orientation of the fracture network. A good reservoir permeability of more than around 10^{-13} m^2 is crucial for the existence of a geothermal reservoir, since it enables convective heat transport.

The **thermal properties** of a geothermal reservoir and its surrounding rocks determine the thermal regime and the amount of heat that can be extracted. The most important property is **thermal conductivity**, which describes the ability of a rock to conduct heat. Thermal rock properties are mostly influenced by the mineralogical composition, but also the porosity and the pore filling (water, air) have a high impact on thermal rock properties. Some rocks exhibit a high thermal conductivity such as granite, others, especially refractory units like mafic rocks, nearly function as a thermal insulator. Sedimentary rocks feature a lower thermal conductivity, whereas igneous rocks tend to have a higher conductivity. For example, a crystalline rock such as granite conducts heat three times better than an unconsolidated rock such as gravel. However, large variations within one rock type are common.

Finally, the **heat production** of rocks, the release of heat due to the decay of radioactive elements, is a steady source of heat within the reservoir and thus influences the potential of a geothermal reservoir. Rocks containing many radioactive elements such as some igneous rocks (granite) have a high heat production.

In difference to thermal and hydraulic properties, which directly influence the potential power output of a geothermal system, **mechanical rock properties** describe the reaction of a rock volume to stress and are important to assess for drilling and reservoir stimulation. When exposed to stress, a rock mass experiences a change in volume or shape, called deformation. Deformation can either be elastic (reversible) or manifest itself in the formation of fractures and cracks (brittle deformation). The latter is called brittleness. It is defined as a property of materials that break without or with only little plastic flow taking place (Yarali and Soyer 2011). Brittleness directly influences the drillability of rocks (e.g. Yarali and Soyer 2011) and strongly differs between sedimentary and crystalline rocks. Reason for this is that mechanical rock properties are, among others, determined by the mineralogical composition of a rock. Quartz, for example, is a mineral found in most crystalline rocks and is very hard in its crystalline form. In contrast, clay is a very soft mineral and is often found in sedimentary rocks. Consequently, rocks are often divided into "soft" and "hard" rocks from a mechanical point of view. More information on mechanical rock properties can, for example, be found in Wittke (1990). Mechanical rock properties also influence seismicity as explained in Sect. 1.4.

1.4   Stress Field and Seismicity

The condition of rocks in the crust results from various stresses and deformation events caused by the past and present **stress field**. For geothermal site selection, it is important to have knowledge about the present stress field, since a change in stress due to hydraulic stimulation can lead to **seismicity**, which is one of the major risks for deep geothermal applications. It is defined as a measure encompassing earthquake occurrence, mechanism and magnitude at a given geographical location (Stacey and Davis 2008). Natural seismicity is a result of stress build up due to tectonic forces and release along fault zones, which are preexisting zones of weakness in the crust (e.g. the San Andreas fault near San Francisco). The Earth's crust is stressed due to tectonic movements. These tectonic forces result in an increasing stress over time, which is suddenly released if the stress exceeds the threshold strength of a fault. Then, the fault ruptures and the accumulated strain energy is released in part as seismic waves, resulting in an earthquake. The threshold strength depends on the fracture roughness and the mechanical properties of the involved rocks. In case the threshold is exceeded due to human activities such as the hydraulic stimulation of geothermal reservoirs, it is spoken of **induced seismicity**.

Since promising locations for geothermal reservoirs are often close to fault zones due to increased permeability, it is crucial to have knowledge of the initial stress state in the subsurface in order to determine the reactivation potential of faults (e.g. Buijze et al. 2019). The present-day **stress regime** in the crust helps to identify regions that are critically stressed. World stress maps (Heidbach et al. 2009) display the orientation of the maximum horizontal stress in the crust, which is decisive for the stability criterion of faults. Seismicity maps show the history and location of seismic events in a region (e.g. National Seismic Monitoring Center, Benz 2017).

To describe the strength of seismic events, different **seismic scales** are used. Seismic magnitude scales describe the overall strength of a seismic event, i.e. the energy released. They are commonly derived from seismograms, which record the amplitude and kind of seismic waves arriving at a certain distance of the earthquake epicenter. The most known magnitude scale is the **Richter magnitude scale** (M_L) (Richter 1935), which is calculated based on the logarithm of the amplitude of seismic waves (e.g. Wallace 1990). The logarithmic scale ranges from 1 (not felt) to greater than 9 (total destruction). Table 1.1 gives an overview of some Richter magnitudes and how they are perceived. In practical applications, however, the moment magnitude scale (Mw, Hanks and Kanamori 1979) is the most established magnitude scale. It

Table 1.1 Excerpt of categories from the Richter magnitude scale (based on Richter 1935)

Magnitude	Description	Earthquake effects
1–1.9	Micro	Rarely felt, but recorded by seismographs
2–3.9	Minor	Often felt by people, rarely causing damage
4–4.9	Light	Felt by most people, shaking of indoor objects
5–5.9	Moderate	Felt by everyone, may cause damage to poorly constructed buildings
6.0–6.9	Strong	Felt several hundred km away from epicenter, damage to well-built structures possible
>7	Major/Great	Damage at most buildings up to total destruction and permanent changes in topography

additionally takes physical properties, such as the rigidity and fault displacement, into account.

In contrast to seismic magnitude scales, seismic intensity scales describe the intensity of ground shaking and how strong it damages a certain area. An example is the **European Macroseismic Scale (EMS)** (Grünthal 1998), categorized from one (not felt) to twelve (completely devastating).

1.5 Classification of Geothermal Systems

All mechanisms and properties described in the previous chapters—temperature gradient, heat flow, rock properties—determine the occurrence and quality of a potential geothermal reservoir. A **geothermal reservoir** is defined as an accumulation of hot fluid within a large volume of porous or fractured rock that can be extracted to the surface. The geothermal reservoir is one component of a geothermal system. Together with a heat source, heat transfer mechanism, heat trap, fluid source, fluid pathways, and fluid trap, it provides the conditions for the accumulation of a geothermal resource, i.e. an accumulation of heat that is reasonably economically extractable (Harvey and Beardsmore 2014).

Geothermal reservoirs can be classified from an economic or a geological point of view. Economic classifications use, for example, the temperature level (Lee 2001) or the enthalpy, both based on reservoir fluid temperatures. The latter classifies geothermal resources into low- (below 100 °C), medium- (100–180 °C), and high-enthalpy (higher than 180 °C) fields. Low-enthalpy resources are used for direct heat use, e.g. for district heating, industrial or agricultural use often in combination with heat pumps. Medium- to high-enthalpy resources allow for power generation using different kinds of geothermal power plants (Sigfússon and Uihlein 2015). Other geothermal reservoir classification schemes can be found in Zarrouk and McLean (2019).

Since the accumulation of a geothermal reservoir is directly linked to the geological setting, this section first introduces the geologically based geothermal reservoir classification scheme of Moeck (2014). Subsequently, we distinguish between conventional or hydrothermal systems and enhanced or petrothermal geothermal systems (EGS), a distinction that is based on the hydraulic properties of the reservoir.

1.5.1 Geothermal Play Types

The occurrence of **favorable geothermal sites** is bound to specific tectonic settings and directly linked to the geodynamic context (Huenges and Ledru 2011). Based on such geological controls, geothermal systems can be classified into geothermal play types. Following Moeck (2014), a geothermal play type is defined according to the plate tectonic setting, the hydrogeology (matrix or fracture porosity) and the nature of the heat source (magmatic or non-magmatic). Building up on this, geothermal plays

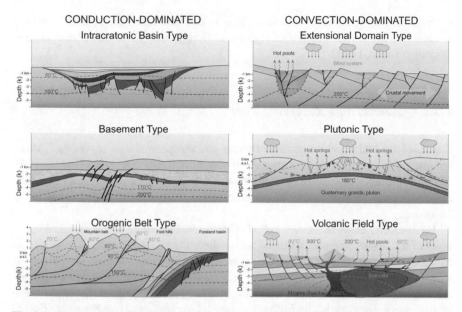

Fig. 1.1 Schematic visualization of geothermal play types. Reproduced from Moeck (2014) with permission from Elsevier and from Harvey and Beardsmore (2014) with permission from the International Geothermal Association

can be distinguished by their dominating type of heat transfer regime (convection or conduction).

Convection-dominated geothermal systems (Fig. 1.1, right) occur in regions of active tectonics or volcanism, for example, in extensional domains. Areas of convection-dominated systems generally feature a high geothermal gradient as well as natural fluid flow. The reason for heat transport is a heat source or elevated heat flow. The proximity to the heat source results in the convection of thermal fluids and thus in the transport of heat from depth to surface. A high permeability is crucial for convection to take place. The following convection-dominated systems are distinguished:

- Volcanic field type: near magmatic arcs, mid-ocean ridges, hot spots (e.g. Iceland).
- Plutonic type: in or close to young orogens with recent plutonism (e.g. Lardarello, Italy).
- Extensional domain type: in back-arc extensional or pull-apart basins, intracontinental rift (e.g. Pannonian Basin, Rhine Graben).

Conduction-dominated geothermal systems (Fig. 1.1, left) are commonly located in passive plate tectonic settings with no recent tectonic or magmatic activity and have an average geothermal gradient. Consequently, these type of resources mostly host low-to-medium temperature resources and are located in greater depths compared to convection-dominated systems. The following types are distinguished:

- Intracratonic basin type: in inactive intracratonic rift basins or passive margin basins (e.g. North German Basin).

- Orogenic belt type: in fold-and-thrust belts or foreland basins (e.g. Molasse Basin).
- Basement type: in intrusions or regions with high heat-producing elements (e.g. Cooper Basin, USA).

1.5.2 Geothermal Reservoir Types

Based on the hydraulic properties of rocks and the presence or absence of water, geothermal plays can host three classes of geothermal reservoir types (Breede et al. 2015): hydrothermal systems, petrothermal systems, and hot sedimentary aquifers. Figure 1.2 gives an overview of the different reservoir types, which we explain in the following.

Hydrothermal systems, also called conventional geothermal systems, are reservoirs of hot water, steam or a mix of both, where the naturally occurring formation water acts as the heat exchange medium. Therefore, the presence of fluid pathways, i.e. good hydraulic conductivity, plays a decisive role in enabling a constant recharge of water (Muffler and Cataldi 1978). The standard approach for exploiting a hydrothermal reservoir is a hydrothermal doublet. It consists of two boreholes drilled into the hydrothermal reservoir: a hot production borehole for extracting the hot fluid and a cold injection borehole for re-injecting the fluid to the subsurface after heat has been transferred via a heat exchanger to a secondary circuit at the surface (Schintgen 2015). An example for this type of geothermal system is Lardarello (Tuscany, Italy).

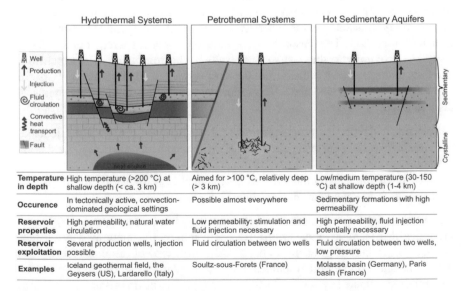

	Hydrothermal Systems	Petrothermal Systems	Hot Sedimentary Aquifers
Temperature in depth	High temperature (>200 °C) at shallow depth (< ca. 3 km)	Aimed for >100 °C, relatively deep (> 3 km)	Low/medium temperature (30-150 °C) at shallow depth (1-4 km)
Occurence	In tectonically active, convection-dominated geological settings	Possible almost everywhere	Sedimentary formations with high permeability
Reservoir properties	High permeability, natural water circulation	Low permeability: stimulation and fluid injection necessary	High permeability, fluid injection potentially necessary
Reservoir exploitation	Several production wells, injection possible	Fluid circulation between two wells	Fluid circulation between two wells, low pressure
Examples	Iceland geothermal field, the Geysers (US), Lardarello (Italy)	Soultz-sous-Forets (France)	Molasse basin (Germany), Paris basin (France)

Fig. 1.2 Schematic visualization of different geothermal reservoir types. Figures modified from Buijze et al. (2019), used under CC BY 4.0 (https://creativecommons.org/licenses/by/4.0/)

As the occurrence of hydrothermal systems is bound to the presence of hot water in permeable rocks, they are the least abundant geothermal reservoir type.

Petrothermal systems, also called unconventional or enhanced geothermal systems (EGS), are volumes of hot rock where natural permeability is too low for economic use. They require artificial stimulation for increasing permeability and accessing the heat stored in the rocks (Breede et al. 2013; Huenges 2010). This can be done, for example, by the injection of pressurized fluids to form new fractures, following the example of oil and gas industry to increase production rate in sedimentary rocks. More information about the EGS technique is given in Sect. 3.4. In petrothermal systems, the hydraulic stimulation increases the size of the subsurface heat exchanger and the injected fluids function as the heat exchanger medium. Petrothermal reservoirs fall in the category of unconventional geothermal resources. Since the EGS technique allows any volume of hot rock in the subsurface to qualify as an artificial geothermal reservoir, petrothermal systems are the most abundant geothermal reservoir type. It is estimated that over 85% of the overall geothermal energy resources are located in petrothermal reservoirs (Richter 2017).

Hot Sedimentary Aquifers belong to unconventional geothermal resources and are aquifers in permeable sedimentary rock formations. Where permeability is not high enough, it can be enhanced using EGS techniques. Compared to hydrothermal resources, they are generally located in greater depth and achieve lower temperature levels. Similar to petrothermal resources, water is circulated between two wells, but at much lower pressure. An example for this type of geothermal reservoirs is the Molasse Basin in Munich (Germany).

References

Benz H (2017) Building a national seismic monitoring center: NEIC from 2000 to the present. Seismolog Res Lett 88(2B):457–461. https://doi.org/10.1785/0220170034

Breede K, Dzebisashvili K, Liu X, Falcone G (2013) A systematic review of enhanced (or engineered) geothermal systems: past, present and future. Geotherm Energy 1(1):4. https://doi.org/10.1186/2195-9706-1-4

Breede K, Dzebisashvili K, Falcone G (2015) Overcoming challenges in the classification of deep geothermal potential. Geotherm Energy Sci 3(1):19–39. https://doi.org/10.5194/gtes-3-19-2015

Buijze L, van Bijsterveldt L, Cremer H, Paap B, Veldkamp H, Wassing BB, van Wees JD, van Yperen GC, ter Heege JH (2019) Review of induced seismicity in geothermal systems worldwide and implications for geothermal systems in the Netherlands. Neth J Geosci 98. https://doi.org/10.1017/njg.2019.6

Grünthal G (1998) European macroseismic scale 1998. Technical report, European Seismological Commission (ESC)

Hanks TC, Kanamori H (1979) A moment magnitude scale. J Geophys Res: Solid Earth 84(B5):2348–2350. https://doi.org/10.1029/JB084iB05p02348

Harvey C, Beardsmore G (eds) (2014) Best practice guide for geothermal exploration. IGA service GmbH, Bochum. https://documents1.worldbank.org/curated/en/190071480069890732/pdf/110532-Geothermal-Exploration-Best-Practices-2nd-Edition-FINAL.pdf. Accessed 19 Oct 2021

Heidbach O, Tingay M, Barth A, Reinecker J, Kurfess D, Mueller B (2009) The 2008 database release of the World Stress Map project. https://doi.org/10.1594/GFZ.WSM.REL2008

Huenges E (2010) Deployment of enhanced geothermal systems plants and CO_2 mitigation (chap 8). In: Geothermal energy systems. Wiley, New York, pp 423–428. https://doi.org/10.1002/9783527630479.ch8

Huenges E, Ledru P (2011) Geothermal energy systems: exploration, development, and utilization. Wiley, New York

Lee KC (2001) Classification of geothermal resources by exergy. Geothermics 30(4):431–442. https://doi.org/10.1016/S0375-6505(00)00056-0

Moeck IS (2014) Catalog of geothermal play types based on geologic controls. Renew Sustain Energy Rev 37:867–882. https://doi.org/10.1016/j.rser.2014.05.032

Muffler P, Cataldi R (1978) Methods for regional assessment of geothermal resources. Geothermics 7(2):53–89. https://doi.org/10.1016/0375-6505(78)90002-0

Pasquale V, Verdoya M, Chiozzi P (2017) Geothermics: heat flow in the lithosphere, 2nd edn. SpringerBriefs in Earth Sciences. Springer International Publishing, Berlin. https://doi.org/10.1007/978-3-319-52084-1

Pollack HN, Hurter SJ, Johnson JR (1993) Heat flow from the Earth's interior: analysis of the global data set. Rev Geophys 31(3):267–280

Richter CF (1935) An instrumental earthquake magnitude scale. Bull Seismolog Soc Amer 25(1):1–32

Richter M (2017) Summary of New Drilling Technologies. Technical report, International Energy Agency - Geothermal (IEA Geothermal). http://iea-gia.org/wp-content/uploads/2014/10/IEA-Geothermal-Drilling-Technologies.pdf. Accessed 25 Oct 2021

Schintgen T (2015) Exploration for deep geothermal reservoirs in Luxembourg and the surroundings - perspectives of geothermal energy use. Geotherm Energy 3(1). https://doi.org/10.1186/s40517-015-0028-2

Schön JH (2015) Physical properties of rocks: fundamentals and principles of petrophysics. Elsevier

Sigfússon B, Uihlein A (2015) 2015 JRC geothermal energy status report - technology, market and economic aspects of geothermal energy in Europe. Technical report, Institute for energy and transport (Joint Reserch Centre), Publications Office of the European Union. https://doi.org/10.2790/959587

Stacey FD, Davis PM (2008) Physics of the earth. Cambridge University Press, Cambridge

Stober I, Bucher K (2020) Geothermie, 3rd edn. Springer Spektrum. https://doi.org/10.1007/978-3-662-60940-8

Toth A (2016) The geothermal atlas of Hungary. Hungarian energy and public utility regulatory authority, pp 125–133

Wallace RE (1990) The San Andreas fault system, California: an overview of the history, geology, geomorphology, geophysics, and seismology of the most well known plate-tectonic boundary in the World. Government Printing Office, U.S

Wittke W (1990) Rock mechanics: theory and applications with case histories. Springer, Berlin. https://doi.org/10.1007/978-3-642-88109-1

Yarali O, Soyer E (2011) The effect of mechanical rock properties and brittleness on drillability. Sci Res Essays 6(5):1077–1088. https://doi.org/10.5897/SRE10.1004

Zarrouk SJ, McLean K (2019) Chapter 2 - Geothermal systems. In: Zarrouk SJ, McLean K (eds) Geothermal well test analysis. Academic, Cambridge, pp 13–38. https://doi.org/10.1016/B978-0-12-814946-1.00002-5

Chapter 2
Geothermal Potential Across Europe

Abstract This chapter introduces the geothermal conditions in Europe and explains the distribution of geothermally favorable regions. From a geological perspective, Europe is split along an intracontinental suture zone extending from the North Sea to the Black Sea. Geothermal conditions, heat flow, and temperature gradients are higher southwest of the suture zone than northeast of it, which is related to the different ages of the crust in these regions. Consequently, positive geothermal anomalies occur mainly south of the suture zone in regions of active tectonic deformation, for example, in the Rhine Graben or the Pannonian Basin, both of which are known for their hydrothermal reservoirs. However, most geothermal reservoirs in Europe are intracratonic basin-type reservoirs located both southwest and northeast of the suture zone. Due to a high potential in geothermal development, we discuss two of them in detail, the Danish and the Polish Basins. We also give a more detailed overview of the Pannonian Basin with above-average geothermal conditions.

Keywords Geothermal potential · Heat flow · Geothermal reservoir distribution · Poland · Sweden · Hungary

2.1 Geothermal Conditions Across Europe

In Chap. 1, we introduced heat flow and temperature gradient as important factors to assess the geothermal conditions of a region, and seismicity as a major risk for geothermal projects. Here, we explain how these three quantities are distributed across Europe. Since these factors are linked to crustal age and tectonic activity, we first give a brief overview of the general geological setting in Europe.

The European continent is divided in two parts of fundamentally different crustal age: The East European Platform and the Baltic shield of Precambrian age (around 2000 Mio. years) in the northeast and Phanerozoic units (younger than 541 Mio. years) in the southwest (Fig. 2.1). The crustal boundary in between is called **Trans-European Suture Zone (TESZ)** and extends from the North Sea to the Black Sea. It crosses the northern part of Scania in Sweden as well as Poland. Due to different researchers discovering marks of the crustal boundary in different countries, the

J. Fink et al., *State of the Art in Deep Geothermal Energy in Europe*,
SpringerBriefs in Earth System Sciences,
https://doi.org/10.1007/978-3-030-96870-0_2

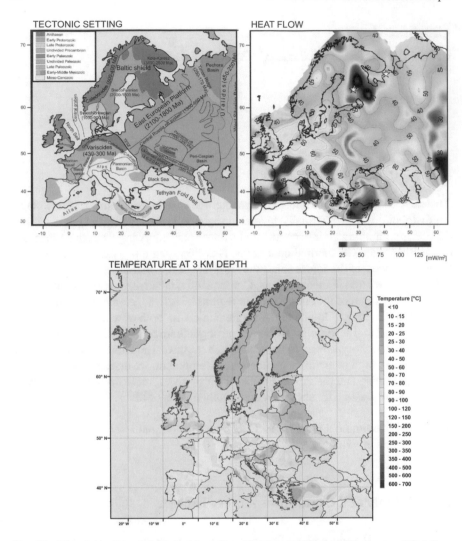

Fig. 2.1 Top: Age of crust (left) and heat flow (right) across Europe. Figures modified from Artemieva et al. (2006). Bottom: Temperature distribution at 3 km depth (Limberger and van Wees 2014, used under CC BY 3.0, https://creativecommons.org/licenses/by/3.0/)

TESZ is termed Sorgenfrei-Tornquist Zone in Sweden and Teisseyre-Tornquist Zone in Poland. Throughout this book, we will consistently use the term TESZ. West of the TESZ, tectonic and magmatic activity is still taking place, manifested, for example, in the formation of the Alps or in volcanism in Italy. In contrast, towards the east, the continent is quiet and stable and no recent tectonic activity has taken place.

There is a direct link between crustal age on the one hand, and both heat flow and temperature gradient on the other hand. The older the crust, the more it has cooled. As a consequence, the TESZ separates two distinct regions in Europe considering the thermal regime. This is highlighted by **heat flow models**, such as the one shown in Fig. 2.1. Heat flow ranges from 25–40 mW m^{-2} in the Precambrian platform to 80–100 mW m^{-2} in the Alpine system. Highest heat flow values of 80–100 mW m^{-2} are reached in the southern Forealps and the Rhine Graben.

Temperature slices at depth show a similar trend (Fig. 2.1). They are important tools to estimate the **geothermal potential** of a region: the closer a high-temperature layer is to the surface, the higher the potential. The temperature distribution in 3 km depth shows, similar to heat flux, a pronounced difference between northeastern and southwestern Europe that follows the TESZ. On the entire European continent, the average temperature at a depth of 5 km is 111 °C, while in the Baltic Shield it is only about 40 °C (Limberger and van Wees 2013). However, the south of Sweden is an exception, making this region interesting from a geothermal perspective, which is why we set a focus on it in Sect. 2.2. One of the regions that reach particular high temperatures at shallow depth (100 °C in 2 km depth) is the Pannonian Basin, which extends over the entire area of Hungary. Consequently, details about the Pannonian Basin will be given in Sect. 2.4.

However, Fig. 2.1 also shows that there is no uniform distribution of temperature and heat flow throughout the two crustal units. Positive heat flow and temperature anomalies are linked to **recent tectonic activity** in southwestern Europe. Figure 2.2 shows the location of the latest active tectonic structures of Tertiary age. Structures such as the Upper Rhine Graben (where European EGS demonstration projects are located) and the Massif Central belong to the Central European Rift System (ECRIS). Figure 2.2 also displays the location of **earthquakes** with a magnitude stronger than 3.0 on the Richter scale. In the Precambrian units earthquakes are rare. Close to active tectonic structures, earthquakes are strongly accumulated. In addition to past earthquakes, world stress databases, such as the one published by Heidbach et al. (2009), help identify regions of active tectonic deformation and basements that are under critically stressed conditions. The previously described regional variations in geology and tectonic regime all relate to different **geothermal play types**. Figure 2.3 shows the distribution of geothermal play types across Europe as well as selected locations of geothermal reservoir types. Geothermal reservoirs are especially located west of the TESZ. Exceptionally high-temperature resources in Europe are located in the volcanic field play type in Iceland as well as in the plutonic play type, for example, in Lardarello, Italy.

However, resources in Europe are dominated by the intracratonic basin type (e.g. Norwegian-Danish Basin, Polish Basin), followed by the extensional domain type (e.g. Pannonian Basin). In the following, we will detail on the geothermal conditions of three countries: Sweden and Poland as representatives for intracratonic basin types, and Hungary, lying in the center of the extensional domain type Pannonian Basin.

2.2 South Sweden

The geology of Sweden is dominated by the Fennoscandian shield, an old and thus stable cratonic region with a very low geothermal gradient (15–16 °C km^{-1}). This makes the conditions for deep geothermal energy usage in Sweden generally less favorable, since very high drilling depths are required to reach the necessary temperature level.

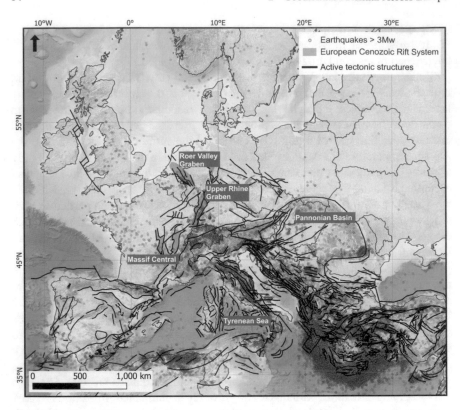

Fig. 2.2 Active tectonic regions in Europe. The Central European Rift system is marked in blue, and purple lines show active tectonic structures. Earthquakes with a Magnitude ≥ 3 Mw highlight regions of natural seismicity. Data from Basili et al. (2013), Ziegler and Dèzes (2006), Cloetingh et al. (2010), the European-Mediterranean Earthquake Catalogue (EMEC) (Grünthal and Wahlström 2012), and EMODnet Bathymetry Consortium (2020)

Thus, most of the area of Sweden is not considered as a geothermal reservoir at all. However, one exception is Scania in the south of Sweden, where Malmö and Lund are located. They are located in the eastern extension of the Danish Basin (Fig. 2.4), a sedimentary basin filled with deposits of Mesozoic age. In this intracratonic basin play type, the average geothermal gradient amounts 28–32 °C km^{-1} (Erlström et al. 2018). Scania lies in the transition between the Fennoscandian shield in the north and younger tectonic regimes in the south, very close to the TESZ. The transition to the older crustal units is marked around 20 km north of Malmö, where the Romeleåsen fault zone crops out (Fig. 2.4).

Hydrothermal and hot sedimentary aquifers in the sedimentary basin-fill succession of the Øresund Basin (part of the Danish Basin) were analyzed by Erlström et al. (2018). Numerous aquifers in different permeable formations, such as the Bunter Sandstone, reach formation temperatures higher than 60 °C in a depth of up to 4 km as well as high natural outflow rates. The average temperature gradient in the

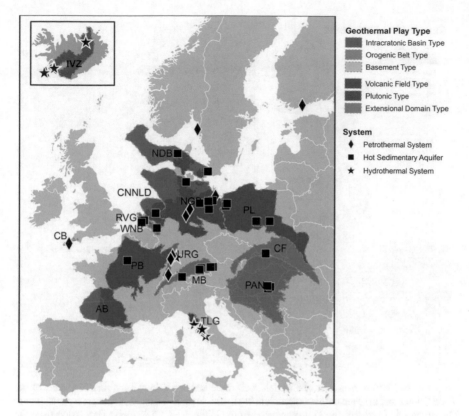

Fig. 2.3 Overview of geothermal plays and a selection of geothermal systems across Europe. Modified from Buijze et al. (2019), used under CC BY 4.0 (https://creativecommons.org/licenses/by/4.0/) CB: Cornubian batholith, CF: Carpathian Mountains and Foredeep, IVZ: Iceland Volcanic Zones, MB: Molasse Basin, NGB: North German Basin, NDB: Norwegian-Danish Basin, PAN: Pannonian Basin, PL: Polish Lowlands, RVG: Ruhr Valley Graben, TLG: Tuscany-Latium Geothermal Area, URG: Upper Rhine Graben, WNB: West Netherlands Basin, AB: Aquitaine Basin

sedimentary succession amounts 20–30 °C km^{-1}, however with strong fluctuations. For example, within the DG-1 well close to Lund, a gradient of only 14 °C km^{-1} was measured. In that well, the transition to the crystalline basement was met in around 2000 m depth, where the gradient increases to 22 °C km^{-1} (cross section in Fig. 2.4; Erlström et al. 2018).

Formation temperatures higher than 100 °C are reached in the **petrothermal reservoirs** in the crystalline basement, buried below the sedimentary succession. Rosberg and Erlström (2019) analyzed their potential for EGS by compiling data from the DGE-1 borehole, which was drilled for geothermal research purposes close to Lund (Appendix 2). The borehole shows that the basement is dominated by gneiss and granite. Their outcrop analogues are highly fractured. These fractures are linked to deep-seated fault zones (e.g. Rosberg and Erlström 2019, marked in Fig. 2.4). Due to repeated reactivation of these fault zones, resulting in damage in the surrounding rock

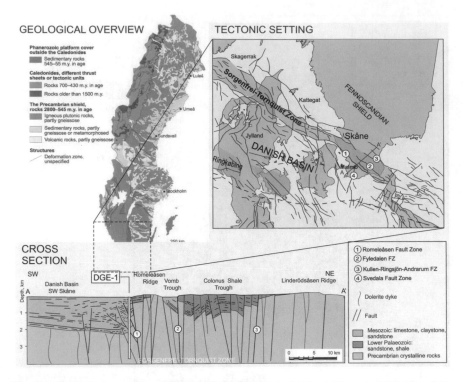

Fig. 2.4 Geological overview of Sweden. Left: Sweden is dominated by very old rocks of the Caledonides and the Precambrian crystalline basement. Only in the south, sedimentary rocks cover the crystalline basement (blue). Reproduced from Gehlin et al. (2019) with permission from the Geological Survey of Sweden. Right: Tectonic setting of south Sweden in the Danish Basin with location of main fault zones and the TESZ, here termed Sorgenfrei-Tornquist Zone, and cross-section location. The cross section illustrates the transition from the sedimentary succession in the Danish Basin to mostly crystalline rocks towards the NE. The location of the DGE-1 well close to Lund is also marked. Modified from Rosberg and Erlström (2019), used under CC BY 4.0 (https://creativccommons.org/licenses/by/4.0/)

volume, it is predicted that the crystalline basement under Scania is hydraulically conductive, especially close to fault zones such as the Romeleåsen fault (Rosberg and Erlström 2019).

However, exact predictions about the basement properties are difficult, since only few wells reach the crystalline basement (Lassen and Thybo 2012; Erlström et al. 2018). Reflection seismic images originally made for oil and gas exploration do not yet provide good images of the basement and need to be re-calibrated for depth. This is a challenge for two reasons: first, the thick sequence of sedimentary successions on top of the basement reduces the signal-to-noise ratio in depth. Second, the structural complexity of the basement is difficult to interpret (Lassen and Thybo 2012). This highlights the need for more data, such as further geophysical surveys and additional drilling, to better characterize the permeability of the basement.

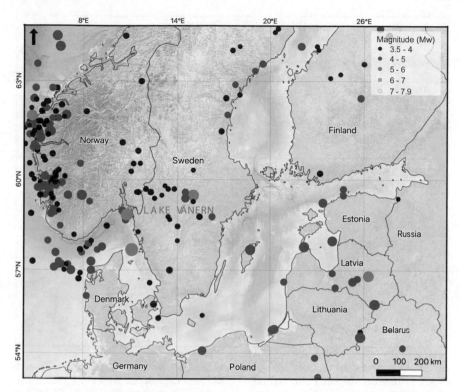

Fig. 2.5 Earthquakes with a Magnitude ≥ 3 Mw in south Sweden. Database is the European-Mediterranean Earthquake Catalogue (EMEC), Grünthal and Wahlström (2012), and EMODnet Bathymetry Consortium (2020)

Regional seismicity in Scania is low, due to its location between the stable and old Fennoscandian shield and the younger crust in the southwest (Fig. 2.5). However, in 2008, a "moderately strong" earthquake with a magnitude of 4.3 on Richter scale occurred in Scania, about 60 km east of Malmö. Its epicenter is located in the TESZ, but it could not be linked to a certain fault zone (Voss et al. 2009).

Tectonic activity in southern Sweden is generally concentrated around Lake Vänern and associated with stress release within the existing fault network and crustal boundaries (Lund et al. 2016). Within individual fault blocks, seismicity is expected to be low.

2.3 Poland

The area of Poland extends over four main European structural units: the South is dominated by the Carpathian Mountain Range, which is part of the Alpine System. The western and eastern parts of Poland are divided by the TESZ, so that parts of

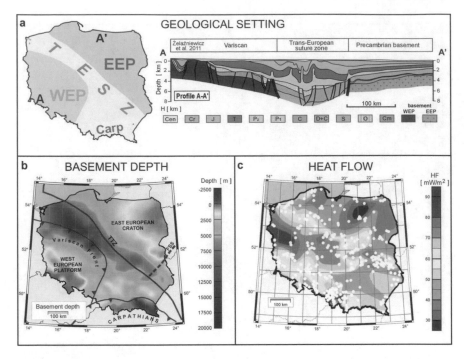

Fig. 2.6 Overview of the structural setting and geothermal conditions in Poland. **a**, left: Poland is located on the East European Platform (EEP) of Precambrian age and the West European Platform (WEP) of Phanerozoic age, separated by the Trans-European Suture Zone (TESZ). The Carpathians (Carp) as part of the alpine system are located in the south. **a**, right: Cross section from A to A' showing the different structural settings throughout the three zones. Black lines are faults. Sediment thickness is highest in the depression along the TESZ. **b**: Depth of the basement below Poland. The basement is deeply buried below thick sediments along the TESZ and also at the Carpathian mountain range. **c**: Heat flow map. White dots indicate the well location of temperature logs. All figures modified from Grad and Polkowski (2016), used under CC BY (https://creativecommons. org/licenses/)

the country are characterized by the Precambrian East European Platform and parts by the Paleozoic units of Central Europe (Fig. 2.6, Sowiżdżał and Kaczmarczyk 2016). Where the TESZ crosses Poland, a complicated deep-reaching fault network is present (Grad and Polkowski 2016). This network was partially reactivated in later stages and led to the accumulation of thick sedimentary units, burying the basement deeply beneath (Fig. 2.6, lower right). Figure 2.6 shows the extent of the four units as well as a cross section, illustrating the transition between Paleozoic and Precambrian Europe. This structural setting causes a major change in the thermal regime throughout the four regions. Heat flow is higher in the southwest (40–70 mW m^{-2}) and decreases towards the northeast (40 mW m^{-2}), where the old Precambrian basement units become more abundant (Fig. 2.6, lower right).

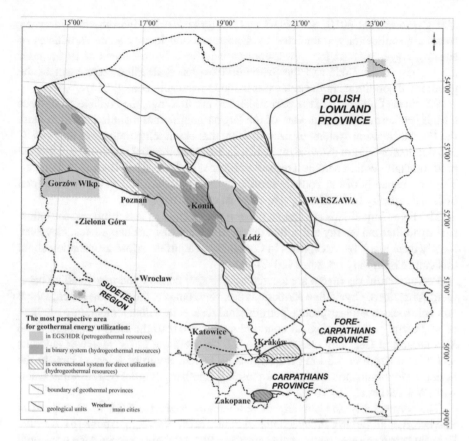

Fig. 2.7 Most prospective areas for hydrothermal (red-shaded area) and petrothermal (light blue areas) energy utilization in Poland (from Sowiżdżał 2018)

In general, Poland mostly hosts **low-temperature geothermal resources**. Sowiżdżał (2018, 2016) give an overview of hydrothermal and petrothermal resources in Poland, which we briefly summarize in the following and which is shown in the map in Fig. 2.7.

The vast majority of **Poland's hydrothermal resources** is located in Permo-Mesozoic sedimentary formations, of which the majority is located in the Polish Lowlands. Thereof, especially the conditions around the Polish Trough are favorable for geothermal energy usage (Fig. 2.7). Sowiżdżał (2018) distinguishes four geothermal provinces: The Polish Lowlands, the Sudetes region, the Carpathians and the Fore-Carpathians (Fig. 2.7).

Largest parts of the country are covered by the Polish Lowlands, comprising the Polish sedimentary basin in which the Polish Trough extends over the central and northern part. The Polish Lowlands host most of the country's geothermal resources and many areas are used for heating and bathing purposes, with temperatures above

100 °C (Sowiżdżał 2010; Sowiżdżał and Semyrka 2016). Hydrothermal reservoir rocks are sedimentary, dominated by sandstones and locally some carbonates of Mesozoic age, such as the Lower Jurassic aquifers. Temperatures of this aquifer are highest in the Lódź and the Szczecin Troughs. A detailed assessment of the geothermal conditions in the latter was made in Miecznik et al. (2015).

Within the Polish part of the Carpathian Mountain Range, the Podhale Trough has a high geothermal potential. One of the largest geothermal installations in Poland, the Podhale heating system, utilizes aquifers that reach temperatures of up to 95 °C in a depth of 2–2.3 km (Sowiżdżał 2018). These hydrothermal reservoirs are located predominantly within Middle Triassic limestones and overlying Eocene carbonates. The total capacity of the Podhale heating system amounts $40.8\,MW_{th}$, with a total installed capacity of $80.5\,MW_{th}$ (Bujakowski et al. 2016).

The Fore-Carpathians are from a general perspective of lesser interest for industrial hydrothermal energy utilization due to relatively low discharge rates. However, locally there are some exceptions in Cenomanian aquifers. More information about this can be found in Sowiżdżał (2018).

Petrothermal resources are located both in sedimentary and in crystalline basement formations. Crystalline formations are, for example, the Gorzow block, where volcanic rocks are a prospective petrothermal reservoir, with a temperature gradient of $40\ °C\,km^{-1}$. The geological conditions are similar to the EGS project in Groß Schönebeck in Germany (see Appendix 2). Temperature in the reservoir amounts 153 °C in a depth of 4.3 km (Wójcicki et al. 2013). South of Jelenia Gora, in the Sudetes Mountains, the Karkonosze pluton is another potential petrothermal reservoir (Wójcicki et al. 2013).

The permeability and porosity of sedimentary rocks for EGS was analyzed in Sowiżdżał et al. (2016). Prospective locations are the Szczecin Trough, parts of the Fore-Sudetic Region as well as some areas in central Poland (see Fig. 2.7). At a depth of about 3.5 km, sedimentary rocks in the Polish Lowlands loose their reservoir properties and the intergranular permeability transitions to fracture permeability which could be enhanced by stimulation. For more information about EGS in sedimentary rocks in Poland, it is referred to Sowiżdżał (2018).

Poland is a country of low **seismic activity**, stronger earthquakes are rare and have long return periods (Guterch 2015). However, an accumulation of historic earthquake epicenters is located towards the mountainous regions in the Sudetes and Carpathians (Fig. 2.8). Along the TESZ and in northwestern Poland, some isolated earthquakes occurred during the last centuries. Even though the Precambrian Platform was supposed to be aseismic, an earthquake stroke Kaliningrad in 2004 with a Magnitude of 5 Mw and led to damage on buildings (Gregersen et al. 2007; Guterch 2015).

In the area of the TESZ, earthquakes accumulate in the vicinity of the Dolsk fault (Lizurek et al. 2013). Macroseismic activity has been registered locally, which implies shallow epicenters (around 4–5 km, Lizurek et al. 2013).

However, since seismicity in Poland is not a major concern, detailed macroseismic data is missing. Consequently, individual studies are necessary for planned EGS projects, especially in areas of high structural complexity.

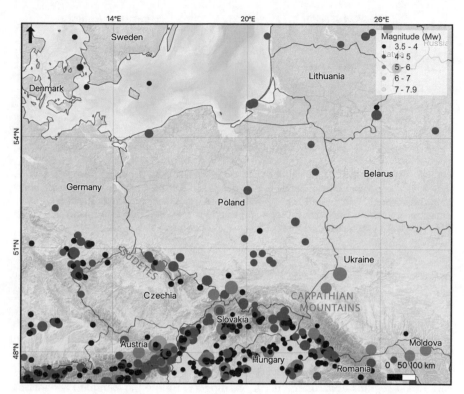

Fig. 2.8 Earthquakes with a magnitude ≥ 3 Mw in Poland. Database is the European-Mediterranean Earthquake Catalogue (EMEC), Grünthal and Wahlström (2012), and EMODnet Bathymetry Consortium (2020)

2.4 Hungary

The conditions for using geothermal energy are particularly advantageous in Hungary (e.g. Rybach et al. 1979; van Wees et al. 2013; Toth 2016). The mean heat flow value is 95 mW m^{-2} and the average geothermal gradient 45 °C km^{-1} (Dovenyi et al. 1983; Toth 2019), which is considerably higher than the continental average. Locally, the gradient can reach values of up to 100 °C km^{-1}, for example, in the Taska Region.

Reason for these conditions is the location in the central part of the **Pannonian Basin**, which is a sediment-filled basin associated with the ongoing orogenesis of the Alps/Himalaya complex (Toth 2016). The basin is encircled by the Carpathian Mountains. As a result of extension, the crust is thinned below the Pannonian Basin, resulting in a positive geothermal anomaly. During basin extension, sediments, reaching a thickness of up to 7 km, were deposited on top of the crystalline basement across the entire area of Hungary. However, the thickness of the sedimentary cover differs locally. For example, in the Little Hungarian Plain, which is the northwestern part of the Pannonian Basin, the cover is made up by up to 7 km thick sequences of

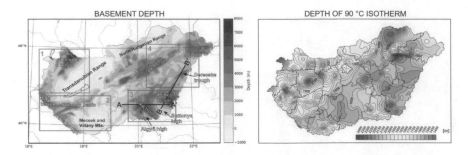

Fig. 2.9 Left: Basement depth and mountain ranges in Hungary. Reprinted from Békési et al. (2018) with permission from Elsevier. See Békési et al. (2018) for more information about the geothermal conditions in the gray boxes as well as the two cross sections. Right: Depth of the 90 °C isotherm. Reproduced from Toth (2016) with permission from the Hungarian Energy and Public Utility Regulatory Authority

sediments of the Danube River (Toth 2016), whereas in the Transdanubian Range, the basement is not covered by sediments at all but crops out at the surface (see Fig. 2.9).

The Geothermal Atlas of Hungary (Toth 2016) assesses and summarizes the geothermal potential, based on data from 1622 thermal wells and 170 abandoned hydrocarbon wells. Resultant isogeothermal maps show that a formation temperature of 50 °C is reached almost everywhere in 700–900 m depth with exception of some karst regions towards the north of Hungary. Below the Zemplen Mountains and the Zolna Hills, where the geothermal gradient is exceptionally high, 70 °C are already reached within a shallow depth of 500–700 m. 90 °C are reached at 1500–1700 m under large areas of the Great Plain and in the Cserehat Region, which is shown in Fig. 2.9 (Toth 2016).

Hungary is well known for its **hydrothermal resources** feeding baths and spas all around the country. Due to the sedimentary basin setting, geothermal reservoir rocks can be distinguished into basement reservoirs and basin-fill reservoirs (Nádor et al. 2019). Basin-fill reservoirs comprise the Pannonian sandstone as well as sedimentary rocks of Miocene age. Basement reservoirs are made up by crystalline basement carbonates and other crystalline basement rocks. Figure 2.10 gives a schematic visualization of the location and depth of these main reservoir formations.

Within the basin-fill, the biggest and currently most exploited hydrothermal reservoir is the late Pannonian sandstone. It is made of barely cemented coarse-grained sand layers with high porosity and permeability and located in a depth range of 1500–2000 m. Temperature ranges between 60 and 90 °C. The formation was subject to extensive hydrogeological and hydrocarbon exploration drillings and its hydrogeology has been studied thoroughly. In 2016, 80% of thermal wells in Hungary extracted water from the late Pannonian sandstone (Toth 2016).

Additionally, hydrothermal basement reservoirs are exploited by deeply buried, interconnected Paleozoic-Mesozoic carbonates. The formation consists of crystalline limestone layers with a high secondary porosity due to extensive faulting, fracturing,

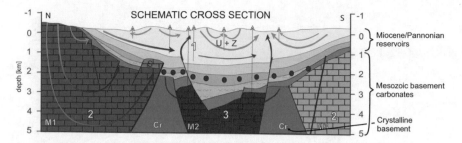

Fig. 2.10 Schematic visualization of geothermal reservoirs in Hungary and fluid flow pathways. Arrows show main fluid pathways. M1-3: Mesozoic basement carbonates. Cr: Crystalline basement. Orange/green: Miocene Basin-fill reservoir. Beige: Pannonian Basin-fill reservoirs. Reprinted from Horváth et al. (2015) with permission from Elsevier

and karstification. On average, the unit lies in a depth of 2000 m or more and reaches temperatures of 120 °C (Nádor et al. 2019). An example for a geothermal well penetrating this unit is in Zalakaros, where water from a depth of 2200–2300 m with a temperature of 99 °C is pumped to the surface. This formation also supplies water to the well-known thermal baths of Budapest (Toth 2016).

Besides the numerous aquifers and hydrothermal reservoirs, the Paleozoic-Mesozoic basement rocks have high potential for **petrothermal energy** usage. They reach temperatures above 200 °C in a depth of 3500–4000 m (Nádor et al. 2019). Some locally occurring granitic and granitoid rocks, especially towards the southeast of the Pannonian Basin, are promising for EGS projects (Garrison et al. 2016; Toth 2019), with high temperatures, a low level of seismicity (extensional regime) and suitable lithologies (granitoid rocks) (Toth 2016). Békési et al. (2018) developed subsurface temperature models of the depth of the basement, to indicate areas where the hot basement rocks are at shallow depth and point to potential locations for EGS installations. Figure 2.9 depicts results of this study. We refer to the publication of Békési et al. (2018) for visualizations of the temperature at the basement top.

Regional seismicity in Hungary is affected by the Pannonian Basin being located between two areas of different seismicity: the Mediterranean and the Eastern European platform. While the latter is almost aseismic, the former is one of the most seismically active regions in the world (Tóth et al. 2006). In contrast to Sweden and Poland, deformation in the Pannonian Basin Region is still ongoing (Tóth et al. 2006). Additionally, convection-dominated geothermal play types are linked to higher natural seismic activity than conduction-dominated systems (Buijze et al. 2019). All in all, seismic activity in Hungary can be considered as moderate, with higher activity towards the surrounding mountain areas, the Carpathians (Fig. 2.11). Majority of earthquakes origin in depths of 6–15 km. However, the exact distribution of earthquakes seems to be diffuse and can hardly be linked to fault lines, which is a consequence of inaccurate or missing data (Tóth et al. 2006). This and the ongoing deformation in the basin require detailed assessments of the risk of seismicity for geothermal projects.

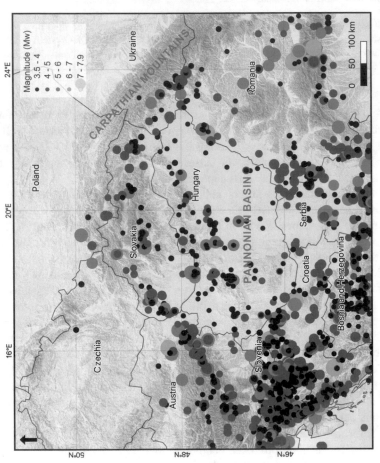

Fig. 2.11 Earthquakes with a magnitude ≥ 3 Mw around the Pannonian Basin. Database is the European-Mediterranean Earthquake Catalogue (EMEC) Grünthal and Wahlström (2012), and EMODnet Bathymetry Consortium (2020)

References

Artemieva IM, Thybo H, Kaban MK (2006) Deep Europe today: geophysical synthesis of the upper mantle structure and lithospheric processes over 3.5 Ga. Geolog Soc Lond Memoirs 32(1):11–41. https://doi.org/10.1144/GSL.MEM.2006.032.01.02

Basili R, Kastelic V, Demircioglu MB, Garcia Moreno D, Nemser ES, Petricca P, Sboras SP, Besana-Ostman GM, Cabral J, Camelbeeck T, Caputo R, Danciu L, Domaç H, Fonseca JFdBD, García-Mayordomo J, Giardini D, Glavatovic B, Gulen L, Ince Y, Pavlides S, Sesetyan K, Tarabusi G, Tiberti MM, Utkucu M, Valensise G, Vanneste K, Vilanova SP, Wössner J (2013) European Database of Seismogenic Faults (EDSF). https://doi.org/10.6092/INGV.IT-SHARE-EDSF

Békési E, Lenkey L, Limberger J, Porkoláb K, Balázs A, Bonté D, Vrijlandt M, Horváth F, Cloetingh S, van Wees JD (2018) Subsurface temperature model of the Hungarian part of the Pannonian Basin. Global Planet Change 171:48–64. https://doi.org/10.1016/j.gloplacha.2017.09.020

Buijze L, van Bijsterveldt L, Cremer H, Paap B, Veldkamp H, Wassing BB, van Wees JD, van Yperen GC, ter Heege JH (2019) Review of induced seismicity in geothermal systems worldwide and implications for geothermal systems in the Netherlands. Neth J Geosci 98. https://doi.org/10.1017/njg.2019.6

Bujakowski W, Tomaszewska B, Miecznik M (2016) The Podhale geothermal reservoir simulation for long-term sustainable production. Renew Energy 99:420–430. https://doi.org/10.1016/j.renene.2016.07.028

Cloetingh S, van Wees JD, Ziegler PA, Lenkey L, Beekman F, Tesauro M, Förster A, Norden B, Kaban M, Hardebol N, Bonté D, Genter A, Guillou-Frottier L, Ter Voorde M, Sokoutis D, Willingshofer E, Cornu T, Worum G (2010) Lithosphere tectonics and thermo-mechanical properties: an integrated modelling approach for enhanced geothermal systems exploration in Europe. Earth-Sci Rev 102(3):159–206. https://doi.org/10.1016/j.earscirev.2010.05.003

Dovenyi P, Horváth F, Liebe P, Gálfi J, Erki I (1983) Geothermal conditions of Hungary. Geofizikai Közlemények 29(1)

EMODnet Bathymetry Consortium (2020) EMODnet Digital Bathymetry (DTM). https://doi.org/10.12770/bb6a87dd-e579-4036-abe1-e649cca9881a

Erlström M, Boldreel LO, Lindström S, Kristensen L, Mathiesen A, Andersen MS, Kamla E, Nielsen LH (2018) Stratigraphy and geothermal assessment of Mesozoic sandstone reservoirs in the Øresund Basin – exemplified by well data and seismic profiles. Bull Geol Soc Den

Garrison GH, Guðlaugsson SÞ, Ádám L, Ingimundarson A, Cladouhos TT, Petty S (2016) The South Hungary enhanced geothermal system (SHEGS) demonstration project. GRC Trans 40:10

Gehlin S, Andersson O, Rosberg JE (2019) Geothermal energy use, country update for Sweden. In: Proceedings world geothermal congress 2020, Reykjavik, Iceland, p 10

Grad M, Polkowski M (2016) Seismic basement in Poland. Int J Earth Sci 105(4):1199–1214. https://doi.org/10.1007/s00531-015-1233-8

Gregersen S, Wiejacz P, Dębski W, Domanski B, Assinovskaya B, Guterch B, Mäntyniemi P, Nikulin V, Pacesa A, Puura V et al (2007) The exceptional earthquakes in Kaliningrad district, Russia on September 21, 2004. Phys Earth Planet Inter 164(1–2):63–74. https://doi.org/10.1016/j.pepi.2007.06.005

Grünthal G, Wahlström R (2012) The European-Mediterranean Earthquake Catalogue (EMEC) for the last millennium. J Seismol 16(3):535–570. https://doi.org/10.1007/s10950-012-9302-y

Guterch B (2015) Seismicity in Poland: updated seismic catalog. In: Guterch B, Kozák J (eds) Studies of historical earthquakes in Southern Poland: Outer Western Carpathian Earthquake of December 3, 1786, and First Macroseismic Maps in 1858–1901, GeoPlanet: earth and planetary sciences. Springer International Publishing, Cham, pp 75–101. https://doi.org/10.1007/978-3-319-15446-6_3

Heidbach O, Tingay M, Barth A, Reinecker J, Kurfess D, Mueller B (2009) The 2008 database release of the World Stress Map Project. https://doi.org/10.1594/GFZ.WSM.REL2008

Horváth F, Musitz B, Balázs A, Végh A, Uhrin A, Nádor A, Koroknai B, Pap N, Tóth T, Wórum G (2015) Evolution of the Pannonian basin and its geothermal resources. Geothermics 53:328–352. https://doi.org/10.1016/j.geothermics.2014.07.009

Lassen A, Thybo H (2012) Neoproterozoic and Palaeozoic evolution of SW Scandinavia based on integrated seismic interpretation. Precambrian Res 204–205:75–104. https://doi.org/10.1016/j.precamres.2012.01.008

Limberger J, van Wees JD (2013) European temperature models in the framework of GEOELEC : linking temperature and heat flow data sets to lithosphere models. In: European geothermal congress 2013, Pisa, Italy, p 9

Limberger J, van Wees JD (2014) 3D subsurface temperature model of Europe for geothermal exploration. In: Conference Proceedings, 76th EAGE Conference and Exhibition 2014. https://doi.org/10.3997/2214-4609.20141657

Lizurek G, Plesiewicz B, Wiejacz P, Wiszniowski J, Trojanowski J (2013) Seismic event near Jarocin (Poland). Acta Geophys 61(1):26–36. https://doi.org/10.2478/s11600-012-0052-6

Lund B, Tryggvason A, Chan N, Högdahl K, Buhcheva D, Bödvarsson R (2016) Understanding intraplate earthquakes in Sweden: the where and why. In: Geophysical research abstracts, Vienna, Austria, vol 18, pp EPSC2016–16441

Miecznik M, Sowiżdżał A, Tomaszewska B, Pająk L (2015) Modelling geothermal conditions in part of the Szczecin Trough – the Chociwel area. Geologos 21(3):187–196. https://doi.org/10.1515/logos-2015-0013

Nádor A, Kujbus A, Tóth A (2019) Geothermal energy use, country update for Hungary. Eur Geotherm Congress 2019:11

Rosberg JE, Erlström M (2019) Evaluation of the Lund deep geothermal exploration project in the Romeleåsen Fault Zone, South Sweden: a case study. Geotherm Energy 7(1). https://doi.org/10.1186/s40517-019-0126-7

Rybach L, et al (1979) Geothermal resources: an introduction with emphasis on low temperature reservoirs. In: Symposium on geothermal energy and its direct uses in the Eastern United States, U.S. Geothermal resources council, special report, vol 5, pp 1–7

Sowiżdżał A (2010) Perspektywy wykorzystania zasobów wód termalnych jury dolnej z regionu niecki szczecińskiej (pó^{3}nocno-zachodnia Polska) w ciep^{3}ownictwie, balneologii i rekreacji [Prospects of use of thermal water resources of Lower Jurassic aquifer in the Szczecin Trough (NW Poland) for space heating and balneology and recreation]. Przegląd Geologiczny 58:9

Sowiżdżał A (2016) Possibilities of petrogeothermal energy resources utilization in central Poland. Appl Ecol Environ Res 14(2):555–574. https://doi.org/10.15666/aeer/1402_555574

Sowiżdżał A (2018) Geothermal energy resources in Poland – overview of the current state of knowledge. Renew Sustain Energy Rev 82:4020–4027. https://doi.org/10.1016/j.rser.2017.10.070

Sowiżdżał A, Kaczmarczyk M (2016) Analysis of thermal parameters of Triassic, Permian and Carboniferous sedimentary rocks in central Poland. Geolog J 51(1):65–76. https://doi.org/10.1002/gj.2608

Sowiżdżał A, Semyrka R (2016) Analyses of permeability and porosity of sedimentary rocks in terms of unconventional geothermal resource explorations in Poland. Geologos 22(2):149–163. https://doi.org/10.1515/logos-2016-0015

Sowiżdżał A, Hajto M, Górecki W (2016) The most prospective areas for geothermal energy utilization for heating and power generation in Poland. In: Proceedings European geothermal congress 2016, p 8

Toth A (2016) The geothermal Atlas of Hungary. Hungarian Energy and Public Utility Regulatory Authority pp 125–133

Toth A (2019) Prospects for geothermal power projects in Hungary. In: ResearchGate

Tóth L, Győri E, Mónus P, Zsíros T (2006) Seismic hazard in the Pannonian Region. In: Pinter N, Gyula G, Weber J, Stein S, Medak D (eds) The Adria Microplate: GPS Geodesy, Tectonics and Hazards. Springer Netherlands, Dordrecht, Nato Science Series: IV: earth and environmental sciences, pp 369–384. https://doi.org/10.1007/1-4020-4235-3_25

Voss PH, Larsen TB, Ottemöller L, Gregersen S (2009) Earthquake in southern Sweden wakes up Denmark on 16 December 2008. GEUS Bull 17:9–12. https://doi.org/10.34194/geusb.v17.5002

van Wees JD, Boxem T, Angelino L, Dumas P (2013) A prospective study on the geothermal potential in the EU. GEOELEC deliverable D2.5, The GEOELEC consortium. http://www.geoelec.eu/wp-content/uploads/2011/09/D-2.5-GEOELEC-prospective-study.pdf. Accessed 22 Oct 2021

Wójcicki A, Sowiżdżał A, Bujakowski W (2013) Evaluation of potential, thermal balance and prospective geological structures for needs of closed geothermal systems (Hot Dry Rocks) in Poland. Warszawa/Kraków (in Polish)

Ziegler PA, Dèzes P (2006) Crustal evolution of Western and Central Europe. Geolog Soc Lond Memoirs 32(1):43–56. https://doi.org/10.1144/GSL.MEM.2006.032.01.03

Chapter 3
Technologies for Deep Geothermal Energy

Abstract Geothermal energy is currently harvested mainly from high-enthalpy resources, i.e. from resources located in regions with favorable geothermal conditions. Most of them are hydrothermal systems and so-called conventional geothermal reservoirs, such as the oldest geothermal field in Lardarello, Italy. However, technological innovation led to an increased exploration and development of low- to medium-enthalpy resources in hot sedimentary aquifers and petrothermal reservoirs. Especially the latter enables geothermal heat use in less favorable regions, thus circumventing geographical limitations. Firstly, this chapter introduces the typical development steps for geothermal projects and associated technologies and methods. Furthermore, it addresses specific requirements for petrothermal reservoirs, providing an overview of unconventional technologies, equipment, and methods, as well as related research and innovation (R&I) activities. For example, it covers innovative drilling technologies suitable for deep hard-rock formations. In addition, the development of petrothermal reservoirs would not be possible without the enhanced geothermal system (EGS) technology that aims at increasing the reservoir permeability by stimulation techniques. We introduce methods of developing petrothermal reservoirs deploying EGS and give an overview of existing large-scale EGS projects in Europe. Here, a main research focus lies on the development of advanced stimulation techniques for minimizing the risk of induced seismicity.

Keywords Geothermal drilling · Exploration · Stimulation · Enhanced Geothermal Systems · Research & Innovation

3.1 Introduction

The deep geothermal sector splits into three main industries: (i) the subsurface industry, (ii) the surface industry, and (iii) the legal and financial service industry (Garabetian 2019). Here, we focus on the subsurface industry, which is the most capital intensive one in geothermal projects. It ranges from exploration to well completion, including reservoir management. In particular, drilling comprises 30–50% of the cost of geothermal projects; for EGS it is even more than 50% of the total cost (Garabetian 2019).

© The Author(s), under exclusive license to Springer Nature Switzerland AG 2022
J. Fink et al., *State of the Art in Deep Geothermal Energy in Europe*,
SpringerBriefs in Earth System Sciences,
https://doi.org/10.1007/978-3-030-96870-0_3

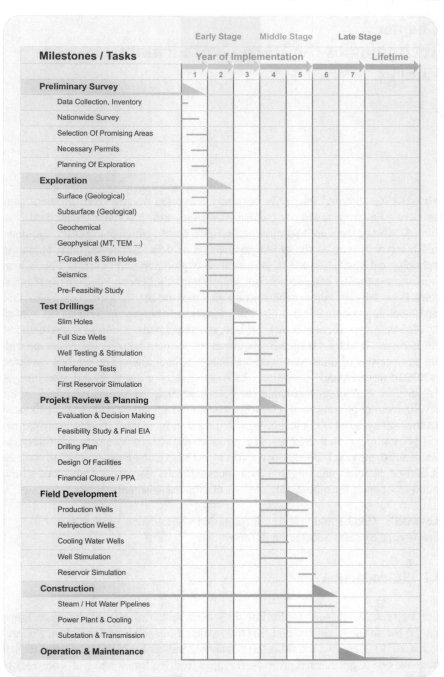

Fig. 3.1 Geothermal project development for a unit of approximately 50 MW. Modified from Gehringer and Loksha (2012), used under CC BY 3.0 IGO (https://creativecommons.org/licenses/by/3.0/igo/)

However, there is more than drilling required for a geothermal project. Its main development stages and tasks are illustrated in Fig. 3.1. Within these development stages, the **European Technology and Innovation Platform on Deep Geothermal** (ETIP-DG, https://etip-dg.eu) identifies five key challenges for deep geothermal in Europe, which they present in their Strategic Research and Innovation Agenda (SRIA) (ETIP-DG 2018):

1. Prediction and assessment of geothermal resources.
2. Resource access and development.
3. Heat and electricity generation and system integration.
4. The shift from R&I to deployment (environmental, regulatory, market, policy, social and human deployment).
5. Knowledge sharing (data harmonization and coordinated organization of data and information, shared research infrastructures).

In addition to the ETIP-DG SRIA, the Deep Geothermal Implementation Working Group (IWG) of the **European Strategic Energy Technology Plan** (SET-Plan, https://www.deepgeothermal-iwg.eu/) defines several R&I activities that refer to direct heat use. Complementary to both, the Geothermal Panel of the ETIP on Renewable Heating and Cooling (RHC ETIP) published its "Strategic Research Innovation Agenda for Geothermal Technologies" (Kujbus et al. 2020) with a focus on geothermal heating.

In Sects. 3.2, 3.3, 3.4, and 3.5, we provide technical details on selected topics from the key challenges 1 and 2 that are related to the subsurface industry. Furthermore, we present both innovative technologies and large-scale projects. To this end, the two Appendices 1 and 2 provide overviews in the form of key points on

1. Recent European technical and non-technical research activities that are related to deep geothermal energy (Appendix 1).
2. Large-scale deep geothermal sites across Europe (Appendix 2).

3.2 Prediction and Assessment of Geothermal Resources

Predicting geothermal resources and selecting areas for detailed exploration both require comprehensive knowledge and a thorough understanding of the subsurface structure and properties. A priori knowledge might be available from former hydrocarbon, geothermal, or mineral exploration. If no or limited prior knowledge is available, the underground geothermal potential is highly uncertain. Geological and geophysical exploration techniques decrease the uncertainty by providing information about the subsurface structure, heat transport processes, petrophysical (e.g. permeability and thermal conductivity), and mechanical rock properties (see Sect. 1.3). Primary risk elements for geothermal resource assessment are temperature (or enthalpy) and transmissivity (i.e. permeability and reservoir thickness). Further important risk elements are reservoir volume, chemistry, and depth (Harvey and Beardsmore 2014). A geothermal exploration program should address these risk elements and aim at providing data for setting up and parameterizing geological and numerical reservoir models. **Numerical reservoir simulations** using well-parameterized models enable sensitivity and uncertainty analysis on the predictions of reservoir temperature, thermal power, and total thermal energy.

Established **exploration methods** and techniques for investigating geothermal resources are described, for example, in the "Best practices guide for geothermal exploration" published by the International Geothermal Association (IGA) and the International Finance Corporation (Harvey and Beardsmore 2014). They range from literature studies, geological mapping, remote sensing, outcrop or reservoir analogue studies, geochemical measurements, geophysical surveys (e.g. gravimetric, geomagnetic, or seismic surveys) to conceptual and numerical modeling (see also Fig. 3.2). The guide also gives an overview of exploration strategies for specific geothermal play types. In addition, we recommend the following sources for further information on the theory and application of the multiple geothermal exploration methods: Bruhn et al. (2010) and Stober and Bucher (2021b).

In the following, we give some more details on geophysical surveys, since they provide a way to image the structure of the Earth's subsurface at depths of up to several km. One of the most established geophysical exploration methods is reflection seismology. Known from petroleum exploration since the 1920s, reflection seismology has become an essential part of exploration programs providing information about the subsurface in a non-invasive manner. In reflection seismology, the structure of the Earth's subsurface is imaged using reflected seismic waves (e.g. Stober and Bucher 2021b). The method requires an artificial seismic source, such as dynamite, and analyzes the seismic waves reflected at geological interfaces. The reflected waves are recorded by geo-phones at the surface by means of their two-way travel time—the time the seismic wave needed to travel from the source to the reflector in the subsurface and back to the surface. The two-way travel time depends on the depth of the reflector and the elastic properties of the subsurface units the wave passed through. Based on the two-way travel time, a seismic section is obtained,

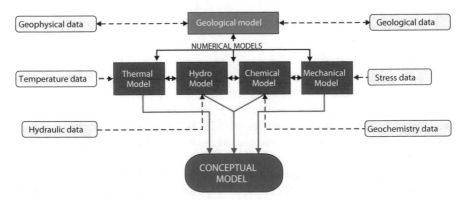

Fig. 3.2 Data input into the conceptual model. Modified from Dezayes and IMAGE SP3 Team (2019) with permission from Chrystel Dezayes

which can be interpreted for geological structures (e.g. layer boundaries, folds, or faults) and fluid reservoirs. Seismic surveys can resolve structures in several km depth, but involve a relatively high logistic effort.

Also electromagnetic surveys have become a frequently used tool for geothermal exploration (e.g. Strack et al. 2008; Spichak and Manzella 2009; Buonasorte et al. 2013). Magnetotellurics (MT) is an electromagnetic method that measures the natural electromagnetic field at the Earth's surface with electrodes and magnetometers (Strack et al. 2008). The natural electromagnetic field reveals the resistivity in the subsurface, which allows for interpeting subsurface structures and fluid pathways from resistivity contrasts. MT is suited for exploring geothermal reservoirs, because higher temperatures and often associated high salinity pore fluids and rock alterations result in decreased rock resistivity. Thus, it is especially applied for exploring hydrothermal reservoirs or volcanic field-type resources (see Sects. 1.5.2 and 1.5.1). MT surveys measure over a wide frequency range (10^{-3}–10^4 Hz), allowing measurements to great depth (Buonasorte et al. 2013). In addition, the survey is less expensive than seismic surveys and logistically practical. However, the measurement can be impaired by anthropogenic noise, such as power lines or railways (Spichak and Manzella 2009).

Continuous R&I efforts aim at improving established methods, such as seismics, or at developing innovative approaches for facilitating exploration and for lowering exploration risk. An overview of advances regarding exploration methods is provided, for example, by the U.S GeoVision Report on exploration (Doughty et al. 2018, pp. 76) or by van Wees et al. (2017). Exploration advancements include, for instance, remote sensing techniques for identifying and mapping thermal anomalies. Satellite thermal imaging (Foley et al. 2016), hyperspectral imaging (Kratt 2011), and Light Detection and Ranging (LiDAR) for mapping fault zones at the surface are just a few examples of methods that have been applied already. As an innovative non-invasive geophysical technique, an ambient-noise tomography of the Greater Geneva Basin has been used in a geothermal exploration context (Planès et al. 2020). Doughty et al. (2018) classify different innovative exploration methods based on their degree of success or of use. Slimhole diamond drilling with real-time temperature logs is an innovative, invasive exploration technique that has proven to be very efficient. For the non-invasive technique, the LiDAR has shown a direct correlation with targeted geothermal resources. The newest technique which has been considered as "game changing" is the combination of repeated microgravity and satellite InSAR. Both methods are combined with a real-time modeling that couples thermal, hydraulic, and chemical processes for directly evaluating permeability or porosity changes in the reservoir. Van Wees et al. (2017) focus on the exploration of high-temperature resources, particularly in Iceland. They recommend distributed acoustic sensing (DAS) and new magneto-telluric (MT) methods that were able to collect data remotely using a wireless system, as well as the joint interpretation of multiple geophysical methods.

Following the selection of a specific (promising) area, detailed exploration includes the drilling of one or more **exploration wells**, which is the only way to obtain direct information and material from the subsurface. Consequently, drilling exploration wells reduces the project risk from high to moderate (Gehringer and Loksha 2012). Usually, exploration wells are shallower than the subsequent production wells, but there is the possibility to expand them to production wells in case of success. An exploration well allows for the collection of direct temperature data as well as direct fluid and rock samples from the subsurface, which can be analyzed in the laboratory. Moreover, borehole measurements (logging) provide direct or indirect information on subsurface parameters (e.g. fractures, permeability, chemical composition, fluid content). Rosberg and Erlström (2021) present, for instance, typical logging and laboratory data from the 3.1- and 3.7-km-deep geothermal exploration wells in Malmö and Lund (Sweden). Alternatively to conventional exploration wells, **slimhole wells** provide a cheaper option for obtaining direct subsurface temperature data (e.g. Mackenzie et al. 2017; Adityatama et al. 2020). The diameter of a slimhole well's production casing is typically between 7 and 17.8 cm (2 3/4–7 in.), whereas conventional well production casings have a diameter ranging from 24.45 cm (standard) to 34 cm (large) (9 5/8–13 3/8 in.) (Mackenzie et al. 2017). When drilled into the reservoir, a slimhole well provides the reservoir temperature and can be used for productivity testing. The standard slimhole drilling technique is rotary drilling with tricone bits. Despite the small diameter, coring is also possible, providing rock samples for further laboratory analysis. Slimhole wells are drilled into depths of up to 2000 m and typically cost 25–40% less compared to conventional wells (Mackenzie et al. 2017). Section 3.3.1 provides more information on geothermal drilling.

All activities during the early project stage, i.e. preliminary survey, exploration, and test drilling (Fig. 3.1), aim at constructing a **conceptual model of the geothermal reservoir** (Fig. 3.2). The conceptual model enables the understanding of the reservoir and thus provides the base for making decisions regarding the next project stages. All data that flow into the conceptual model contribute to risk reduction, identification of uncertainties and their quantification. The analysis of the conceptual model is carried out by analytical calculations and numerical simulations. For instance, the heat stored in the reservoir rock, which is a main factor for the production potential, is calculated from values defined in the conceptual model, such as the reservoir thickness and the reservoir temperature. **Numerical simulations** based on the conceptual model allow, among other things, testing of production scenarios, prediction of long-term reservoir behavior, and quantification of uncertainties. Thus, they assist the exploration process by defining or minimizing risks (i.e. uncertainty), for instance, by stochastic simulations (e.g. Vogt et al. 2012, 2013; Deb et al. 2020) or by predicting optimal locations for exploration wells (Seidler et al. 2016).

Moreover, complex **thermo-hydro-mechanical and chemical (THM/C) interactions** must be investigated in order to conduct a thorough analysis of the geothermal reservoir and fully predict its behavior, thus minimizing risks. This is seen most prominently, but not exclusively, in EGS. For instance, chemical processes such as mineral precipitation alter the mechanical rock properties, or a change in temperature causes a change in fluid density and viscosity, which in turn affect the hydraulic behavior. Laboratory experiments improve the understanding of coupled THM/C processes and enable numerical THM/C modeling on reservoir scale, thus allowing to understand and predict processes in the reservoir. The governing equations for heat transport, fluid flow, deformation, and reactive transport are collected into a strongly nonlinear system solved via numerical approaches (e.g. Kolditz et al. 2012; Tao et al. 2019; Vallier et al. 2019). This way, THM/C modeling presents an important method for addressing numerous questions related to an EGS project, such as the prediction of induced seismicity or the estimation of the lifetime of a geothermal reservoir. THM/C modeling can be applied in multiple stages of the geothermal project after exploration. It can optimize geothermal exploitation by predicting the productivity of the EGS depending on design parameters (e.g. location of the wells, injection flow rates, and temperature). Besides, THM/C modeling allows for assessing the risks of the EGS project (Chap. 4) such as induced seismicity and scaling.

The primary goals of research and innovation (R&I) in the field of resource prediction and assessment are to reduce the cost of exploration technologies and to increase the likelihood of success in exploring deep geothermal resources. For the latter, innovative technologies comprising surface measurements, remote sensing, and imaging and monitoring at surface and borehole level have to be developed (ETIP-DG 2018). According to the Strategic Research and Innovation Report (ETIP-DG 2018), the main topics for R&I in the field of geothermal resources prediction and assessment are

1. Improved exploration prior to drilling (e.g. van Wees et al. 2017).
2. Advanced investigation and monitoring technology (e.g. Shook and Suzuki 2017).
3. Exploration workflows—conceptual models, reservoir characterization, performance, and decision models (e.g. Bilić et al. 2020).
4. Exploration catalogues—reservoir analogues, rock properties, and model constraints (e.g. Dalla Santa et al. 2020, for shallow geothermal applications).
5. Assessing resource potential (e.g. Ciriaco et al. 2020).
6. Beyond conventional resources (e.g. Bertani et al. 2018, EGS: Sect. 3.4 and Appendix 2).

3.3 Resource Access and Development

After the successful assessment of a geothermal resource, the reservoir has to be accessed through boreholes and to be developed for producing and using the heat from the subsurface. We report on both conventional technologies and innovations with a focus on drilling methods and reservoir enhancement. Technologies for accessing and developing deep geothermal resources depend largely on the geological conditions and characteristics of the specific geothermal system (see Sect. 1.5). Typical geological settings vary from igneous (e.g. Iceland), metamorphic (e.g. Lardarello, Italy), magmatic (e.g. Soultz-sous-Forêts, France), and sedimentary (e.g. Groß Schönebeck, Germany) settings, resulting in different physical and mechanical rock properties (see Sect. 1.3). Additionally, geothermal reservoirs vary in temperature conditions and depths (i.e. pressure conditions), and reservoir fluids differ in their chemical composition. All these factors determine the choice of the suitable drilling technology and equipment, monitoring and measurement approaches as well as technology and material for well completion. In addition, a low reservoir permeability may require stimulation methods for creating or enhancing a heat exchanger in low permeable deep rocks (EGS, see Sect. 3.4).

Different **decision-making tools** exist (e.g. Rajšl et al. 2019) for taking all these factors into account. A multi-criteria decision-making tool developed by Raos et al. (2019) considers geological information, technical aspects, energy and heat prices as well as environmental criteria. Input criteria can be weighted by importance. As a result, the tool returns optimal operation plans, the levelized cost of energy (LCOE), cost of investment and environmental impact. The tool was successfully applied to five different geothermal locations in the Pannonian Basin (Bilić et al. 2020).

However, there is a large potential of improving the existing methods and technologies in order to reduce costs and risks. According to the SRIA (ETIP-DG 2018), the main R&I topics related to resource access and development are

1. Advancement towards robot drilling technologies (e.g. Macpherson et al. 2013; Bello et al. 2015).
2. Rapid penetration rate technologies (e.g. Jamali et al. 2019, GEO-DRILL project (Appendix 1)).
3. Green drilling fluids (e.g. Ekeinde et al. 2019; Al-Hameedi et al. 2020).
4. Reliable materials for casing and cementing (e.g. Thorbjornsson et al. 2017).
5. Monitoring and logging while drilling (including "looking ahead of the bit") (e.g. Islam and Hossain 2021).
6. High-temperature electronics for geothermal wells (e.g. Ásmundsson et al. 2014; Riches and Johnston 2015).
7. Effective and safe technologies for enhancing energy extraction (e.g. Sutra et al. 2017; Gischig et al. 2020; Zhang and Zhao 2020).
8. Total re-injection and greener power plants (e.g. Manente et al. 2019; Niknam et al. 2020).
9. Reducing corrosion and scaling and optimizing equipment and component lifetime (e.g. Nogara and Zarrouk 2018; Azari et al. 2020).
10. Efficient resource development (e.g. Verma and Torres-Encarnacion 2018).
11. Enhanced production pumps (e.g. Kullick and Hackl 2017).

Several of these topics are currently addressed by EU-funded research projects listed in Appendix 1. In Sects. 3.3.1 and 3.4, we give more details about the state of the art and innovations for two key technologies for accessing deep geothermal resources: drilling technologies and EGS.

3.3.1 Geothermal Drilling

Conventional drilling technology originates from the petroleum industry. Low-to-medium enthalpy reservoirs in sedimentary settings are similar to hydrocarbon reservoirs. Thus, standard petroleum drilling technology can be used. Yet, the well completion should differ, as geothermal well completions aim at a high-volume flow rate (Dumas et al. 2013). However, the greatest geothermal potential lies in deep hard-rock formations. Thus, deep geothermal drilling requires equipment and methods suitable for high formation temperatures, hard rocks, high pressures, and chemically aggressive formation fluids. Typically, there is an exponential increase of drilling costs with increasing depth, which makes deep drilling very expensive. Yet, drilling deep might be inevitable for reaching desired temperatures. At the same time, the profit margin for geothermal energy is much less compared to the petroleum industry. Therefore, faster and less expensive drilling methods are needed. Generally, a fast rate of penetration (ROP) (e.g. Baujard et al. 2017b) and minimal tripping times are desired for minimizing drilling costs. Some common risks during drilling are "stuck pipe issues, mud losses, casing running problems, wellbore instability associated with stuck pipe and casing running, cementation problems, drilling through hard formation" (Teodoriu and Cheuffa 2011). We refer to Sperber et al. (2010) for a comprehensive overview and descriptions of conventional drilling equipment, completion technologies, drilling processes, as well as associated risks. In the following, we explain some selected conventional and innovative technologies and components of the drilling process.

Drilling technology comprises all processes and equipment that support a drilling method. The latter is the actual process of destroying the rock and transporting the drill cuttings, i.e. crushed or scratched rock material, to the surface. Five common drilling technologies are casing drilling, coiled tubing drilling, underbalanced drilling, managed pressure drilling and slim hole drilling (Teodoriu and Cheuffa 2011). With the casing drilling technology, the well is drilled and cased simultaneously, using the casing as the drill string that is rotated with a top-drive at the surface. Casings are hollow pipes that are used for stabilizing the borehole and for preventing the loss of drilling fluid. Casing drilling is rather used for shallow wells than for deep hard-rock formations. Coiled tubing drilling uses the conventional drilling assembly with a downhole motor. In underbalanced drilling, the wellbore fluid pressure is lower than the natural pore pressure gradient, resulting in an increased ROP. Managed pressure drilling aims at controlling the pressure profile in the well to guarantee the stability of the wellbore (Teodoriu and Cheuffa 2011). Advances in well completion, the casing, materials, or techniques for handling the well pressure can contribute to a safer, faster, and less expensive drilling.

Regarding established drilling technology, Dumas et al. (2013) provide a database of deep drilling companies and a best practice handbook for geothermal drilling. Figure 3.3 illustrates a typical geothermal borehole construction plan. Polsky et al. (2008) emphasize factors that have to be considered for the well construction process: Required drilling tools, preferred drilling practice, well trajectory, well branching, number of casing strings and casing diameters required, and well completion. These factors are influenced by the geothermal reservoir type (hydrothermal or EGS), resource depth, and lithology. In case of an EGS, the stimulation methodology, the production strategy, and the intervention strategy are additional influencing factors. For example, depending on the planned stimulation methodology, the well trajectory could be vertical or deviated (Fig. 3.3). Depending on the production and intervention strategies, the well completion might be open hole or slotted liners, or a specialized completion for selectively choking or restricting individual zones. This way the

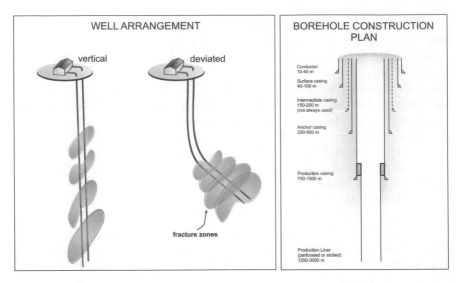

Fig. 3.3 Left: Schematic illustration of vertical and deviated well arrangements with vertically oriented fracture zones in an EGS. Reproduced from Polsky et al. (2008) with permission from Yarom Polsky. Right: Schematic well completion plan for a typical geothermal well (data from Dumas et al. 2013)

injection can be controlled for preventing thermal breakthrough (Sect. 3.5) or short circuiting between injection and production well or the loss of injected fluid (Polsky et al. 2008).

Important advances in the drilling technology are the development of **directional drilling** and **multilateral wells** (e.g. Islam and Hossain 2021). Directional drilling allows for deviating from a vertical drill path by using a steerable guided drill together with electronic monitoring of the drill path. A deviation from the vertical path may be desired, for instance, for increasing the borehole length within the reservoir formation or for deviating from a fixed well location at the surface, e.g. in a densely populated urban area. Similarly, by using the directional drilling technology multilateral wells can be drilled, deviating from one vertical borehole.

Table 3.1 gives an overview of the particular technologies for destroying the rock, namely, the **drilling methods**. The conventional drilling method is **rotary drilling**, where a drill bit, to which a downward force is applied, is turned by rotating the entire drill string. The drill bit has to be exchanged several times during the drilling process because of wearing, which results in tripping time. Realistically, the ROP is between 1 and 2 m per hour and the service life is approximately 50 to 60 hours for conventional tools (Richter 2017). The drill cuttings are transported to the surface within a fluid that circulates down the drill string, through the bit, and up the annular space between the hole and the drill string (Teodoriu and Cheuffa 2011). Drilling fluids are necessary for controlling the borehole pressure, cooling and lubricating the drilling bit, removing cuttings, and carrying them to the surface. Moreover, the drilling fluid transmits hydraulic energy to the measurement and logging tools and

Table 3.1 Overview of selected drilling methods (adapted from Sigfússon and Uihlein 2015a)

Technology	Advantages	Disadvantages
Rotary drilling, roller cone bits	Compatible with existing hardware	Low Rate of Penetration (ROP)
Rotary drilling, drag bits	Compatible with existing hardware	Low ROP
Rotary drilling, hybrid	Compatible with existing hardware, up to four times faster than roller cone bits	Low ROP but faster than rotary drilling with drag bits
Hammer drilling	Compatible with existing hardware, suitable for wide shallow conductor and surface casings	Uncommon in deep formations, no efficiency in soft formations
Jetting	Up to 100 times higher ROP than diamond bits in hard formations	Large pressure losses in deep wells, special equipment, 10–15 times higher energy demand than rotary drilling
Laser, ablation	No information available	Immature technology, complicated transport of beam into deep wells
Laser, spallation	Less energy intensive than ablation	Immature technology, complicated transport of beam into deep wells
Spallation	Can work in deep hydrothermal environment	Immature technology, 10–15 times higher energy demand than rotary drilling
Plasma bit	Potentially 4 times faster ROP than conventional drilling	Immature technology, pilot system exists
Millimeter wave	Rate of penetration should be 10–15 m per hour and should not decrease with increasing depth	Has only been demonstrated in the laboratory, energy intensive
High-voltage electrical impulses	Potentially 5–10 times faster ROP than conventional drilling	Immature technology, high current and power transmission from surface to the bottom, special insulators needed, ROP may decrease with depth

the bit. The drilling fluid can be gas, gas/water, water-based mud, oil-based mud, or synthetic fluids. Gas (usually compressed air or nitrogen) is used, in particular, as drilling fluid in two cases. First, in underbalanced drilling, where the hydrostatic head of the drilling fluid needs to be lower than the pressure of the drilled formations. Second, in dry hard rock, where stability is not a problem (Sperber et al. 2010).

Another common mechanical drilling method is **percussion drilling**, where a heavy cutting or hammering bit is repeatedly lifted and dropped for destroying the rock. In hard-rock formations, down-the-hole (**DTH**) **hammer drilling** is applied. With DTH drilling, the hammer is located inside the hole just behind the drill bit, instead of being located on top of the drill string. DTH hammer drilling has an increased ROP, especially when combined with rotary drilling, but it is insufficient for soft rocks (Teodoriu and Cheuffa 2011). Conventionally, the DTH hammer is powered by air. The very deep production wells in the Otaniemi geothermal project (Espoo, Finland, see Appendix 2) were drilled with the air DTH hammer technology up to 4.4 km depth. The first borehole was continued using both water hydraulic hammer drilling and conventional rotary drilling (St1 2021). The applicability of air-driven DTH hydraulic hammers is limited in water-bearing formations because air can lift water only down to certain depths. In the Malmö exploration well, for instance, formation water influx into the borehole prevented the application of air-driven hammer drilling after around 90 m of drilling into the crystalline basement. Thus, drilling was continued with conventional rotary drilling with an ROP varying between 1 and 4 m per hour, whereas the ROP for the DTH was 15 m per hour (Rosberg and Erlström 2021). Therefore, research and development focuses on the development of hydraulically powered DTH hammers that are powered by water. However, for an environmental-friendly and effective drilling method, DTH fluid/mud hammers need to be developed, which have the ability to use drilling mud as the working fluid. This technology is not operative and market ready yet (e.g. Richter 2017; Wittig et al. 2015). Currently, the DTH fluid/mud hammering method is one main research topic in the EU-funded project GEO-DRILL (Appendix 1).

Besides new hammer drilling approaches, experimental and **innovative drilling methods** comprise chemical drilling methods, jetting, thermal drilling, direct stream, electrical plasma, and millimeter waves for destroying the rock (Table 3.1). Jet drilling uses the drilling mud energy for destroying the rock but is not yet the adequate technology for deep geothermal drilling according to Sigfússon and Uihlein (2015a). The International Energy Agency's geothermal section (IEA Geothermal) provides a comprehensive summary of new drilling technologies (Richter 2017). Recently, the **combined thermo-mechanical drilling method** (CTMD) has been tested in the laboratory and shows promising results in first prototype field applications (Rossi et al. 2020a, b, c, d). Here, flame jets provide thermal assistance for the conventional rotary drilling process. High temperature and high heating rates induce cracks in the rock resulting in faster penetration and less bit wear.

Another recently explored advanced drilling method is **Plasma-Pulse for Geo-Drilling** (PPGD). With this contact-less method, electric pulses induce an electric arc (plasma channel) inside the rock creating tensile fractures. PPGD reduces drilling efforts because it is energy efficient and produces almost no tool wear. It is currently applied at laboratory scale and investigated by numerical simulations in research at ETH Zürich (NN 2021; Ezzat et al. 2021). An important requirement for all new drilling methods is their compatibility with existing hardware (surface and downhole) in order to be applicable to field conditions (Teodoriu and Cheuffa 2011).

An integral part of the whole drilling process is the monitoring of the drilling progress and the wellbore. To this end, **downhole logging** tools are used to measure, for example, the borehole diameter and shape. In addition, various geophysical tools collect formation evaluation data, such as petrophysical analysis, geomechanical analysis, or reservoir fluid characterization. There is equipment for either logging while drilling (LWD) or for downhole logging of the entire borehole or specific depth intervals after the drilling equipment has been removed from the well (e.g. Zarrouk and McLean 2019). In special cases, e.g. during directional drilling, sensors can be installed at the drill head. This is called **monitoring while drilling** (MWD). Additionally, MWD supports the steering for deviated wells by measuring the inclination (e.g. Sperber et al. 2010). We refer to Zarrouk and McLean (2019) for a full description of geothermal well test analysis during and after drilling.

Borehole logging and monitoring equipment has been developed for hydrocarbon reservoirs whose temperatures usually do not exceed 150 °C. However, much higher temperatures can occur in geothermal reservoirs especially for the volcanic play type, i.e. their temperatures exceed the operational range of conventional logging equipment. As a result, development of high-temperature electronics and materials that, for example, resist corrosion in high-temperature and aggressive geothermal environments is an ongoing R&I topic. The European research projects Geo-Coat and Geo-Drill address this topic (see Appendix 1).

3.4 Enhanced Geothermal Systems (EGS)

After a brief introduction of the principle of Enhanced Geothermal Systems in Chap. 1, this chapter deals in more detail with the EGS technique and describes its history and pilot EGS projects in Europe. The EGS concept aims at increasing the reservoir permeability using stimulation and forcing a circulation of the natural brine in the deep reservoir. From an economic point of view, EGS projects can be divided into different categories. The US research on EGS distinguishes two categories (Wertich et al. 2018):

- **In-field and near-field EGS**: Geothermal sites that represent an extension of an identified hydrothermal system or conventional reservoir. The in-field EGS projects in the US are usually focused on re-using unproductive geothermal well fields—such as the Desert Peak site and its closeby The Bradys site—by reservoir stimulation. Near-field EGS are geothermal sites that are associated with an identified hydrothermal system or hot sedimentary aquifer (see Sect. 1.5.2 and Fig. 1.2), where EGS technology is deployed to, for instance, allow production at shallower depth or increase natural flow rates (e.g. Unterhaching, Germany).
- **Green-field EGS**: Geothermal sites away from any clearly identified hydrothermal system or a prior conventional geothermal site. The green-field EGS is more difficult to build than the in- and near-field EGS. All European deep EGS projects—such as Soultz-sous-Forêts or Rittershoffen—and petrothermal reservoirs classify as green-field EGS (see Sect. 1.5.2 and Fig. 1.2).

3.4.1 Stimulation Techniques

Economic production of geothermal heat or electricity from enhanced geothermal systems (EGS) relies on stimulation treatments. Several stimulation techniques exist to enhance a geothermal reservoir, which means the increase of the reservoir permeability and porosity. By stimulating, the size of the subsurface heat exchanger is increased by creating new fractures, reopening existing fractures, or by increasing the fracture permeability.

The most important technique is **hydraulic stimulation**, where liquid is injected into a reservoir to increase its permeability by a combination of developing new tensile and shear fractures, and tensile opening and shearing of preexisting fractures (Fig. 3.4). Stimulated fractures may stay open naturally through the self-propping effect (see Fig. 3.4) or they have to be kept open artificially by proppants injected together with the stimulation liquid to achieve a permanent permeability enhancement (Gentier et al. 2005; Hofmann et al. 2018). The reservoir enhancement can be enormous. For example, Evans et al. (2005) report a 200-fold increase of the

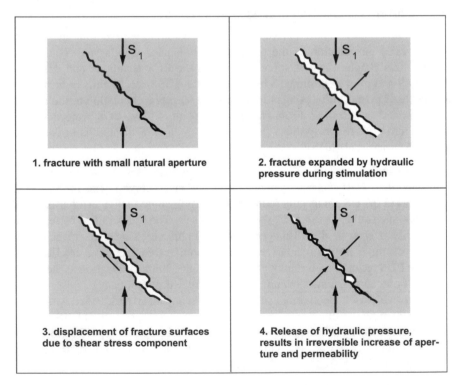

Fig. 3.4 Schematic illustration of the processes within a single fracture during hydraulic stimulation. Pressured fluid injection results in an increased fracture aperture. This effect is also known as "self-propping" (S_1: largest principal stress) (Stober and Bucher 2021a)

reservoir transmissivity after hydraulic stimulation at Soultz-sous-Forêts. However, the enhancement is constrained by the initial fracture distribution and the local stress field, for instance, shearing is more easily triggered for fractures oriented parallel to the local stress field (Pine and Batchelor 1984). Therefore, knowledge about the local stress field and preexisting fractures is paramount for hydraulic stimulation treatments. The success of the hydraulic stimulation in terms of the fracture propagation can be monitored by recording and analyzing the acoustic emissions caused by the pressure build-up and by micro-movements of the fracture (Gaucher et al. 2015). Hydraulic stimulation can jeopardize an EGS project by causing seismic events that can be felt on the surface or even cause damage. Hofmann et al. (2018, 2019) propose an **advanced fluid injection** scheme that aims at mitigating these unwanted events and improving the permeability enhancement process. It involves different types of cyclic injection (Fig. 3.5) and a traffic-light system. The procedure ensures that the condition for reservoir stimulation is approached slowly, since the largest seismic event sometimes occurs during shut-in (Deichmann and Giardini 2009). It also enables to observe the acoustic emissions and to adapt the injection schedule accordingly. Moreover, numerical simulations using coupled hydraulic fracturing simulators can assist the prediction of fracture propagation in order to plan and optimize the stimulation process (e.g. Deb et al. 2021). Aposteanu et al. (2014) estimate the technology readiness level (TRL)[1] of hydraulic stimulation at 6, since the risk of induced seismicity still hinders an application on industrial scale (see Chap. 4).

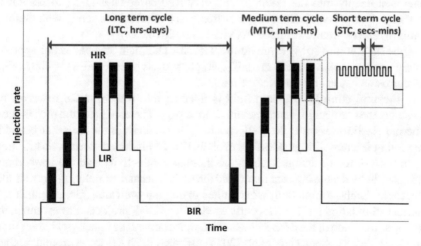

Fig. 3.5 Cyclic fluid injection protocol as presented by Hofmann et al. (2018) for soft stimulation. It shows the long-term (LTC), medium-term (MTC), and short-term (STC) cycles with alternating phases of high injection rate (HIR), low injection rate (LIR), and base injection rate (BIR) (Hofmann et al. 2018, used under CC BY 4.0, http://creativecommons.org/licenses/by/4.0/)

---

[1] The TRL is a measure for characterizing a technology's maturity level. It ranges from TRL 1—lowest level where research is beginning—to TRL 9—highest level when an actual system has been proven in operational environment (e.g. NASA 2021).

Besides hydraulic stimulation, chemical or thermal treatment of a geothermal reservoir can be suitable for increasing its permeability and hence the flow rates. **Chemical stimulation** uses acids for dissolving either minerals that close fissures or pores in the reservoir rock, or mineral deposits within the wells that reduce fluid flow rates. Chemical stimulation has been applied to deep EGS projects during the previous 20 years for enhancing the fracture network particularly in EGS projects (Portier et al. 2007). Sutra et al. (2017) provide a review on chemical stimulation and processes. Chemical stimulation allowed, for instance, the enhancement of the fracture network in granitic geothermal reservoirs such as in the Fjällbacka and Beowawe sites (see also Sect. 3.4.2). Used chemicals are, for instance, formic acid and methanol at the Saint-Gallen site, Switzerland (Sutra et al. 2017). Chemical stimulation has also been applied at Soultz-sous-Forêts, France, where improvements of the productivity were obtained for several wells (Portier et al. 2009a). For example, the GPK-4 well was stimulated with so-called regular mud acid for dissolving hydrothermal minerals from the fractures of the granite. The treatment resulted in a considerable increase in porosity and fracture connectivity and a slight increase in permeability. Thus, the chemical stimulation distinctly increased the productivity. However, it should be considered that acid treatment poses the risks of damaging the well and polluting the environment. Alternatively, environmentally friendly chemicals can be used as seen at the Rittershoffen geothermal site, France (see also Sect. 3.4.2, Vidal et al. 2016). Therefore, the injectivity index—a measure of the performance of injection wells—was increased by a factor of around 1.7 using a biodegradable chelating agent, which penetrated deeply into the reservoir and dissolved carbonates and sulfates. Overall, the geothermal reservoir at Rittershoffen was treated subsequently with thermal, chemical, and hydraulic stimulation (Vidal et al. 2016).

Aposteanu et al. (2014) estimate the TRL for chemical stimulation at approximately 5, since well materials are still being improved in order to resist the scaling and corrosion caused by chemical stimulation.

In **thermal stimulation**, cold fluid is injected into a warm EGS reservoir for increasing the permeability and fracture connectivity. The cooling of the rock induces a thermo-elastic stress near the well, causing micro-cracks in the rock matrix and in preexisting fracture fillings (Griffiths et al. 2017). The induced thermo-elastic stress depends on different factors such as the thermal conductivity and the temperature difference between the injected liquid and the rock. Thermal stimulation has a similar effect as hydraulic stimulation, but is limited to the near-well area (Grant et al. 2013). Thermal stimulation has been recently used in deep EGS projects. For example, the thermal stimulation of the deep Rittershoffen well doubled the injectivity index (Vidal and Genter 2018). Aposteanu et al. (2014) estimate the TRL to be around 5, since this technology has only been demonstrated in few operational environments. More details about reservoir stimulation methods can be found in several textbooks and other publications and references therein (e.g. Huenges and Ledru 2011; Kumari and Ranjith 2019; Stober and Bucher 2021a).

3.4.2 EGS Demonstration Projects

In Europe, there are currently three plants that generate electricity from Enhanced Geothermal Systems (Insheim and Landau, Germany; Soultz-sous-Forêts, France), and one EGS plant that produces heat (Rittershoffen, France). Further plants are under development, such as the Otaniemi plant in Espoo, Finland, or the United Downs Deep Geothermal Power Project in Cornwall, UK (see Appendix 2). Breede et al. (2013) provide a comprehensive overview of the worldwide development of EGS projects until 2013 and a more recent review is provided by Lu (2018).

Usually, **EGS projects in Europe** are located in areas with an above-average geothermal gradient and were chosen for profitability reasons. One of the most promising regions is the Upper Rhine Graben with the highest temperatures in 5 km depth in Central Europe. There, the network of natural fractures inside the granitic basement can be locally activated by stimulation (Lu 2018). Because of these favorable conditions, many EGS sites throughout Europe are located in the Upper Rhine Graben. In the following, we give an overview about selected EGS pilot projects. Additionally, Appendix 2 provides a compilation of European EGS plants as well as selected hydrothermal plants.

The **EGS research and demonstration project in Soultz-sous-Forêts**, France, located in the Rhine Valley, was already started in the 1990s (Baria et al. 1999; Gerard and Kappelmeyer 1991). In 2008, its capacity has been recorded about 1.5 MW$_e$ with a reservoir temperature of about 165 °C (Genter et al. 2010). It includes several, up to 5 km, deep wells intersecting the sedimentary cover over a depth interval of 1.4 km and reaching the deep granitic basement. All wells intersect at least one permeable natural fracture zone. The project brought fundamental insights into the characteristics of natural fractures, fluid geochemistry, temperature and hydraulic properties of deep crystalline rock masses. Extensive data sets have been collected, including mineralogical, petrophysical, borehole logging, passive and active seismic data. The data provide insight into the coupled thermal, hydraulic, mechanical, and geochemical processes that occur during fluid circulation (Genter et al. 2010). The knowledge collected in Soultz-sous-Forêts motivated the development of other promising EGS projects in France, Germany, and Switzerland such as Rittershoffen (Baujard et al. 2017a), Landau (Schindler et al. 2010; Heimlich et al. 2015), and Bruchsal (Meixner et al. 2014; Herzberger et al. 2010).

The **EGS project in Rittershoffen** (France), initiated in 2011, is also located in the Upper Rhine Graben, only 6 km east of Soultz-sous-Forêts. Consequently, its geological setting is very similar to Soultz-sous-Forêts. In Rittershoffen, a geothermal doublet produces heat (hot water at about 170 °C) from the reservoir at the sediment-basement interface utilizing the Rittershoffen fault (Baujard et al. 2017a). The first 2.6 km deep well was drilled at the end of 2012; well testing and subsequent reservoir development operations were realized from January to June 2013. Thermal, chemical, and hydraulic stimulation was performed during this period of time, allowing an increase of productivity. The second 3.2 km deep well was drilled from May to July 2014, and well testing and a circulation test were carried out from August to October

2014. The construction of the 15 km-long heat delivery loop and of the power plant started in February 2015. The geothermal plant went in operation by mid-2016 and is now producing around 25 MW$_{th}$ (Baujard et al. 2017a). The project is considered a successful EGS project and features an innovative geothermal heat plant as well as a special financing scheme (Vidal and Genter 2018).

The **EGS project in Landau** (Germany), located in the northwest of the Rhine Graben, was initiated in 2003. It consists of two wells, 3.3 and 3.1 km deep, drilled from 2005 to 2006 (Schindler et al. 2010). The exploited fractured granite reservoir starts in 2.4 km depth. The reservoir temperature is estimated to be around 160 °C (Hettkamp et al. 2007). The injectivity of this reservoir was increased by hydraulic and chemical stimulation from approximately 0.251 s^{-1} bar^{-1} to more than 11 s^{-1} bar^{-1} (Schindler et al. 2010). The Landau site is generating electricity by producing up to 3 MW$_e$ in 2010 (Schindler et al. 2010).

3.5 Reservoir Operation

The conventional way of producing geothermal heat from a deep reservoir, both hydrothermal and EGS, is via a **geothermal doublet** (see also Fig. 1.2). The doublet consists of one production and one (re-)injection well. The hot fluid is pumped up from the reservoir via the production well. Depending on the surface system, the heat contained is either converted into electrical energy in a power plant or captured with a heat exchanger and used directly, for example by feeding it into a district heating network. Alternatively, a combination of power generation and heat use is possible by using a Combined Heat and Power Plant (CHP, Eyerer et al. 2020).

Power generation requires temperatures of more than 120 °C. Sigfússon and Uihlein (2015a) give a short and illustrative overview of different geothermal power plant designs for electricity production, while Saadat et al. (2010) provide a more technical description of energetic utilization options. Improved and innovative power plant designs aim at enhancing the heat-to-power conversion at low temperatures (60–100 °C), such as the smart mobile Organic Rankine Cycle (ORC) units that are currently investigated in the research project MEET (see Appendix 1). At the same time, surface technologies that allow for a flexible and intelligent response to variations in heat demand must be developed. For instance, the GeoSmart project aims at developing innovations for flexible geothermal systems (see Appendix 1). Regarding heat use, a **cascade usage** concept is possible, where the residual heat is used at decreasing temperature levels for appropriate purposes such as space heating, heating greenhouses, or heating water. Furthermore, the evolution of district heating technologies and the renovation of buildings and district heating systems will gradually allow for lower supply temperatures (IRENA and Aalborg University 2021).

The geothermal reservoir works as a subsurface heat exchanger, where the reinjected fluid is heated up before it reaches the production well. If the fluid does not have enough time to heat up on its way through the reservoir, the heat exchanger system does not work anymore. This is called a thermal breakthrough. Therefore, the placement of injection and production well and the distance in between are critical factors. They require good planning, taking into account the reservoir properties, the conceptual reservoir model and numerical studies (Li et al. 2016). Same holds for the re-injection strategy comprising the completion plan of the injection well (as already described at the end of Sect. 3.3.1) and the injection rates. This step is typically more critical in EGS, because the outcome of the stimulation—that is, the flow pathways between injection and production wells—is more difficult to predict and usually has higher uncertainties than in hydrothermal reservoirs. Sometimes geothermal triplets are used, consisting of two production and one injection well to increase the productivity, for example at the EGS site Soultz-sous-Forêts, France (Portier et al. 2009b, see also Sect. 3.4.2).

As an alternative to the conventional geothermal doublet, a **one-well concept** can be used, where the cold fluid is injected into the subsurface through the inner part of the well (the tubing) and the hot fluid pumped up within the annulus. Such a single-well circulation scheme relies on deploying a very well insulated tubing, i.e. vacuum-insulated tubing, which has high costs. At the same time, the expenses for a second deep well are omitted. The research and demonstration project GeneSys (Hannover, Germany, see Appendix 2) was a pilot project for the one-well concept (Tischner et al. 2010). The project envisaged the circulation of the fluid between two formations. However, the project had to be abandoned due to salt deposition within the well. Currently, the Eden Project (UK) is in the construction phase of a coaxial circulation system within one 4500 m deep well (Eden Geothermal 2021, see Appendix 2).

Another promising innovative technology is the Eavor Loop™ project (http://eavor.com) that received the Ruggero Bertani European Geothermal Innovation Award 2020 (http://egec.org/european-geothermal-innovation-award). The Canadian technology-based energy company Eavor developed a **closed-loop system** for deep geothermal energy that can provide geothermal energy on industrial scale. They suggest to create a closed-loop heat exchanger by joining vertical wells by several horizontal or multilateral wells in the subsurface. Subsequently, they circulate the working fluid through the loop. This system neither relies on subsurface permeability nor does it require artificial stimulation (EGS). In contrast to existing closed systems, where only a small amount of energy is produced from single deep wells, 40 MW and more can be produced according to Eavor. Analytical models by Yuan et al. (2021) evaluate the thermodynamic efficiency of such a closed-loop system that consists of multiple lateral wells. For the analyzed application example, the provided energy production is relatively stable for a time span of 30 years. Their modeling results indicate that thermal conductivity is a crucial parameter for defining the optimal production layer for a closed-loop system. Critical design factors for the system are the optimization of lateral length, spacing and number of horizontal wellbores, as well as fluid velocity. Additionally, further advances in horizontal drilling technology are required to overcome technological difficulties and reduce costs (Yuan et al. 2021). Eavor has realized a prototype installation called Eavor-Lite™ at a pilot site in Alberta, Canada, aiming at demonstrating the technological readiness and its feasibility. The prototype installation depicted in Fig. 3.6 consists of two vertical

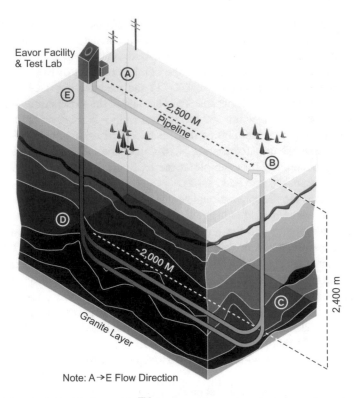

Note: A→E Flow Direction

Fig. 3.6 The principle of the Eavor-Lite™ demonstration project of a deep geothermal closed-loop system with two vertical wells hat are connected in depth by horizontal boreholes. Modified from http://eavor.com/about/technology (accessed 20 May 2021) with permission from Eavor GmbH

boreholes that are joined by two multilateral legs at 2.4 km depth. The technology has not been realized in a commercial project to date. However, a commercial project is intended in Germany (Geretsried, Bavaria) by a joint venture of Eavor and Enex Power Germany GmbH (Eavor 2020).

In conventional hydrothermal or petrothermal reservoirs, the reservoir can be operated not only with the reservoir fluid (brine), but also with other working fluids of different chemical composition and physical properties. For instance, Brown (2000) proposed **CO_2-based EGS**, a concept that uses CO_2 as fracturing and working fluid. Moreover, the concept combines geothermal heat extraction with carbon capture and storage (CCS). The potential of CO_2-based EGS was investigated in several numerical studies (Pruess 2006, 2008; Atrens et al. 2009, 2010) and recently summarized by Wu and Li (2020). Supercritical CO_2 has several advantages over brine. Its substantially lower kinematic viscosity (i.e. greater fluid mobility) improves heat advection within reservoirs. Further, its strongly temperature-dependent density results in a density difference between the injection and production wells that can drive the CO_2 circulation without the need for fluid circulation pumps (so-called thermosiphon effect).

The use of CO_2 as subsurface working fluid is also a novelty for sedimentary basin reservoirs and called direct **CO_2-Plume Geothermal (CPG) system** (Fleming et al. 2020). Here, CO_2 is produced from the geothermal reservoir, expanded in a turbine for generating electricity, cooled, and re-injected into the reservoir. As the CO_2 ultimately remains stored underground, the CPG approach combines CO_2 capture, utilization, and storage (CCUS). Several numerical modeling studies investigated and indicated its potential (Adams et al. 2021, 2015, 2014; Fleming et al. 2020; Garapati et al. 2015; Randolph and Saar 2011). To our knowledge, up to now, neither the CPG nor the CO_2-based EGS approach have been applied to a real reservoir.

Both conventional and these technologies for deep geothermal energy involve certain risks. Considering risks and developing strategies for their mitigation are essential steps of project planning. We provide an overview of the technical as well as non-technical risks of deep geothermal projects and associated mitigation strategies in Chap. 4.

Appendix 1: Selected European Research Projects on Deep Geothermal

In this section, we present short profiles of selected European research projects that are related to deep geothermal energy and EGS. Most projects are funded by the EU's Horizon 2020 program and deal with R&I related to deep geothermal reservoirs and geothermal heating. They are either ongoing or were completed recently. Information was gathered from the respective project websites, which are also listed.

GEODH: Geothermal District Heating

Duration: 2011–2014
Main goals/results:

- Information hub about Geothermal District Heating in Europe
- Increasing awareness of this technology
- Developing strategies for simplification of administrative and regulatory procedures
- Developing innovative financial models
- Training technicians, civil servants, and decision-makers to provide background to approve and support projects

Demonstration sites: 27 case studies across Europe (http://geodh.eu/database)

System/Application: Geothermal District Heating systems
Innovations:

- GeoDH Geographical Information System
- Short comprehensive guides for stakeholders and the public on geothermal district heating in Europe

Project Website: www.geodh.eu.

DESTRESS: Demonstration of Soft Stimulation Treatments of Geothermal Reservoirs

Duration: 2016–2020
Main goals/results:

- Demonstrating methods of EGS
- Adapting and applying recently developed stimulation methods
- Developing best practice guides and workflows
- Delivering specific stimulation technologies for proto-type demonstration
- Providing demonstrators for innovative stimulation treatments

Demonstration sites:

- Bedretto, Switzerland
- Geldinganes, Iceland
- Groß Schönebeck, Germany
- Mezõberény, Hungary
- Rittershoffen and Soultz-sous-Forets, France
- Haute-Sorne, Switzerland (stopped)
- Klaipeda, Lithuania (stopped)
- Pohang, South Korea (stopped)
- Westland, The Netherlands (stopped)

System/Application: EGS in different lithologies
Innovations: Innovative stimulation treatments:

- Combined hydraulic-acidization treatments in sandstones and other rocks
- Cyclic hydraulic and multi-stage stimulation in granites and tight sandstones

Challenges: Several stopped demonstrations for different reasons:

- Chemical stimulation did not result in hydraulic improvement
- Hydraulic stimulation triggered a Mw 5.5 earthquake (in Korea)
- Borehole in the Netherlands (Rotliegend) found too low reservoir permeability, commercial project not possible even with successful near wellbore stimulation
- Project delay due to a legal process initiated by opponents (Switzerland)

Project website: www.destress-h2020.eu/en/home.

PERFORM: Improving Geothermal System Performance Through Collective Knowledge Building and Technology Development

Duration: 2018–2021
Main goals/results: Implementation and evaluation of capabilities to control mineral scaling, particles clogging, corrosion and temperature/ stress-related effects of geothermal flow and injectivity Expected to result in an increase of the energy output by 10–50%.
Demonstration sites: 36 in total, 6 key plants:

- Denmark: Margretheholm, Sønderborg Fjernvarme, Thisted Varmeforsyning
- Germany: Gross Schönebeck
- Netherlands: Honselersdijk, Pijnacker Nootdorp

System/Application: Central European plants with high salinity, high heavy metals concentration, and relatively low temperatures (60–170 °C)
Innovations: Demonstration of new and improved, cost-effective technologies to prevent clogging and corrosion: low-cost cation extraction filters, self-cleaning particle removal appliances, H_2S removal technology, and soft-stimulating injection procedures (thermal and CO_2-injection).
Challenges: Problems and learnings are summarized on the project webpage
Project website: www.geothermperform.eu.

MEET: Multi-sites EGS Demonstration

Duration: 2018–2021
Main goals/results:

- Gathering knowledge of EGS in various geological settings
- Increasing heat production from existing plants and convert oil wells into geothermal wells
- Enhancing heat-to-power conversion at low temperatures (60–90 °C) by using smart mobile Organic Rankine Cycle units
- Providing roadmap of promising sites where EGS solutions for electricity and heat production could be replicated in a near future

Demonstration sites:

- Cazaux, Chaunoy, Condorcet, and Soultz-sous-Forets, France
- Death Valley Analog, USA
- Grásteinn and Krauma, Iceland
- Havelange, Belgium
- United Downs Deep Geothermal Project, UK
- Universitätsenergie Göttingen GmbH, Germany

System/Application: EGS
Innovations: Open Access Decision Support Tool for optimal usage of Geothermal Energy; Demonstration of electricity and thermal power generation from various geological contexts
Project website: www.meet-h2020.com.

DARLINGe: Danube Region Leading Geothermal Energy

Duration: 2017–2019
Main goals/results:

- Increasing the use of geothermal energy
- Advancing stakeholder cooperation
- Delivering data of deep geothermal energy resources at southern part of Pannonian Basin
- Establishing a market-replicable tool box

Demonstration sites: Three cross-border pilot areas: Slovenia-Hungary-Croatia, Serbia-Hungary-Romania, Serbia-Bosnia and Herzegovina

System/Application: Energy-efficient cascade system

Innovations:

- Market-replicable tool box with three complementary modules:
    - an independent indicator-based benchmark evaluation of current uses
    - a decision tree tool to help developers
    - a geological risk mitigation tool to maximize the success rate of a first geothermal well
- Web-map viewer with data from study region

Project website: www.darlinge.eu.

GEOSMART

Duration: 2019–2023

Main goals/results: Combine thermal energy storage with flexible ORC solutions for improving the flexibility and efficiency of geothermal heat and power systems

Demonstration sites:

- Balmatt, Belgium (low enthalpy)
- Kizildere field, Turkey (high enthalpy)

System/Application: Low- and high-enthalpy CHP

Innovations:

- Hybrid cooling system for ORC to prevent efficiency degradation due to seasonal variations
- Scaling reduction system: specially design retention tank, heat exchanger, and recombination with extracted gases

Project website: www.geosmartproject.eu.

GEO-DRILL: Development of Novel and Cost-Effective Drilling Technology for Geothermal Systems

Duration: 2019–2022.

Main goals/results:

- Reducing drilling costs through DTH hammer
- Advancing drill monitoring through low-cost and robust 3D printed sensors
- Improving component life through advanced materials and coatings

System/Application: Deep geothermal with high temperatures
Innovations:

- Robust DTH hammer with high ROP and ability to use drilling mud for improved cuttings transport
- High-fidelity drill monitoring system based on robust 3D printed sensor, cables, compatible in extreme geothermal environments, for fast-wired communication to enable real-time monitoring system
- Economical and efficient methods, materials, and designs in surface engineering to produce high-performance coated surfaces optimized for operating in the aggressive environments of geodrilling
- Integrated tool for sustainability assessment and decision-making

Project website: www.geodrillproject.eu.

GEO-COAT: Development of Novel and Cost-Effective Corrosion Resistant Coatings for High Temperature Geothermal Applications

Duration: 2018–2021
Main goals/results: Developing specialized corrosion- and erosion-resistant coatings optimized for operation in aggressive Geothermal environments
System/Application: Aggressive geothermal environments
Innovations:

- Novel coatings based on selected high entropy alloys and ceramic/metal mixtures
- Knowledge-based system database
- Flow assurance simulator

Project website: www.geo-coat.eu.

GEOHEX: Advanced Material for Cost-Efficient and Enhanced Heat Exchange Performance for Geothermal Applications

Duration: 2019–2022
Main goals/results: Developing material with anti-scaling and anti-corrosion prop-
                      erties
Innovations: Improved heat exchanger materials
Project website: www.geohexproject.eu.

GEOPRO: Accurate Geofluid Properties as Key to Geothermal Process Optimization

Duration: 2019–2022
Main goals/results:

- Understanding and modeling geofluid characteristics
- Improving the accuracy and consistency of key thermodynamic and kinetic input data for increasing the efficiency of plant operations

Demonstration sites: Turkey, Iceland, Germany
Project website: www.geoproproject.eu.

GEOENVI: Tackling the Environmental Concerns for Deploying Geothermal Energy in Europe

Duration: 2018–2021
Main goals/results:

- Mapping environmental impacts and risks and defining
  how environmental footprint is measured and controlled
  in different (European) countries
- Building a harmonized methodology to assess environmental impacts
- Life Cycle approach: Generic database on environmental
  concerns

Demonstration sites:

- Theistareykir, Iceland
- Soultz-sous-Forêts and Rittershoffen, France

- Bagnore 3 and 4 geothermal power plants, Amiata, Italy
- Kizildere, Turkey
- Balmatt, Belgium
- Szeged district heating system, Hungary

Innovations: Simplified Life Cycle Assessment methodology that calculates
                        environmental impacts and benefits in one day
Project website: www.geoenvi.eu.

GEORISK: Developing Geothermal and Renewable Energy Projects by Mitigating Their Risks

Duration: 2018–2021
Main goals/results: Establishing risk insurance for deep geothermal plants all
                        over Europe and in key target third countries
Demonstration sites: Europe, Chile, Mexico, Kenya
Innovations:

- Risk register: list of plausible risks and corresponding de-risking measures
- Risk assessment for different regions

Project website: www.georisk-project.eu.

Appendix 2: Selected Deep Geothermal Sites Across Europe with a Focus on EGS Sites

In the following, we present brief overviews of selected large-scale deep geothermal sites across Europe. The presented projects comprise research facilities and pilot projects, successfully operating commercial plants, as well as abandoned sites. We focus on EGS projects—both petrothermal and hot sedimentary aquifers. In addition, three hydrothermal reservoir projects are given, which are exemplary for many more hydrothermal plants in Europe.

EGS—Petrothermal

Finland: Otaniemi Project, Espoo

Status:	Ongoing (development and construction phase)
Lithology:	Precambrian igneous rocks: Gneiss, amphibolite, granitic intrusions
Wells:	2 wells up to 6000 m
Reservoir:	100–110 °C
Capacity:	Planned: up to 40 MW_{th}, covering up to 10% of Espoo's district heating demand
Purpose:	Commercial plant
Funding:	Energy company St1 and public subsidies
Specialties:	

- World's deepest industrial geothermal energy project
- Hydraulic stimulation phase finished
- Microseismic events during and after stimulation (max. M 1.9)
- Ongoing flow tests in both wells
- Ongoing construction and installation of the above-ground power plant

Source of information: Kukkonen and Pentti (2021), Leonhardt et al. (2021), Hillers et al. (2020), and St1 (2021).

France: Soultz-sous-Forêts

Status:	Ongoing (since 1984)
Lithology:	Granite
Wells:	One well 3600 m deep plus three wells 5000 m deep (2 production wells, 1 injection well)
Reservoir:	165 °C, 20 l s^{-1}
Capacity:	1.5 MW_e, ORC plant since 2008
Purpose:	Research facility and pilot plant
Funding:	€80 Mio. (€30 Mio. EU, €25 Mio. Germany, €25 Mio. France)
Specialties:	Two 3600 m deep wells drilled first, later one well was deepened to 5000 m and two additional boreholes were drilled to 5000 m
Source of information:	Sigfússon and Uihlein (2015b).

France: Rittershoffen

Status: Ongoing (since 2008)

Lithology: Granite
Wells: 2500 m deep
Reservoir: 150–170 °C
Capacity: 24 MW$_{th}$
Purpose: Commercial plant
Specialties: First European EGS providing industrial heat
Source of information: Mouchot et al. (2018).

France: GEOVEN, Strasbourg

Status: Abandoned in 2020
Lithology: Transition between sedimentary cover and granitic base-
                              ment
Wells: 5400 and 6000 m (doublet)
Reservoir: >150 °C
Capacity: Planned: electricity for up to 20000 households, heat for
                              26000 households
Purpose: Commercial plant
Specialties: Seismic events during well testing resulted in the closure
                              of the project, largest event was M3.6.
Source of information: GEOVEN (2021) and Schmittbuhl et al. (2021).

Germany: Bad Urach

Status: Abandoned
Lithology: Metamorphic, Gneiss
Wells: 4445 m (Urach 3) and 2793 m (Urach 4)
Reservoir: Urach 3: 170 °C, 50 l s^{-1} during fracturing test
Capacity: 1 MW$_e$ (planned)
Purpose: Pilot plant
Specialties:

- One of the first EGS projects on pilot scale worldwide
- Urach 4 was planned to drill until 4300 m but geological difficulties, loss of drilling fluid, financial difficulties occurred
- Urach 3: torn off bore rod

Source of information: Sigfússon and Uihlein (2015b).

Germany: Groß-Schönebeck

Status:	Ongoing (since 2000)
Lithology:	Sandstone and andesitic volcanic rocks
Wells:	4309 m (reopened abandoned borehole from gas exploration) and 4400 m
Reservoir:	145 °C, 201 s^{-1}, hydraulic gel proppant and fracturing, chemical fracturing
Capacity:	10 MW$_{th}$, 1 MW$_e$ (ORC) planned
Purpose:	Research facility
Specialties:	In situ geothermal laboratory
Source of information:	Sigfússon and Uihlein (2015b).

Germany: GeneSys Hannover

Status:	Abandoned
Lithology:	Sedimentary (Bunter Sandstone)
Wells:	Horstberg: 3800 m, Hannover: 2900 m
Reservoir:	

- Horstberg: 150 °C, 10–201 s^{-1}
- Hannover: 160 °C, 71 s^{-1} (planned), hydraulic fracturing

Capacity:	Aim: 2 MW$_{th}$ with 25 m^3 h^{-1} at 130 °C for heating the Geozentrum Hannover
Purpose:	Research facility
Funding:	€15 Mio.
Specialties:	

- Pilot project for single-well concept
- 20000 m^3 freshwater have been injected (up to 801 s^{-1})
- Microseismicity in Hannover (M1.8)
- Hannover: Injected freshwater dissolved high amounts of salt, salt deposition occurred during pumping up the hot water
- Horstberg: abandoned gas well used for research purposes

Source of information:	Sigfússon and Uihlein (2015b).

Hungary: South Hungarian EGS Demonstration Project, Battonya

Status:	Ongoing (since 2012), in exploration and planning phase (status 2016)

Lithology: Igneous, Granite
Wells: Planned depth 3500–4000 m, four production and two
                            injection wells planned
Reservoir: Expected: 225 °C at 4000 m depth, 280 kg s^{-1}
Capacity: 8.9 MW$_e$ (planned)
Purpose: Commercial plant
Funding: €39 Mio. from NER300 (Europe), €56 Mio. project costs
Specialties: Planned hydraulic stimulation in a compressional stress
                            field: pressure up to 350 bar, stimulation of multiple frac-
                            ture zones
Source of information: Ádám and Cladouhos (2016) and EU-FIRE (2016).

Sweden: Fjällbacka

Status: Concluded (1984–1995)
Lithology: Granite
Wells: 70–500 m
Reservoir: 16 °C, flow rates of 0.8–1.8 l s^{-1}
Capacity: None
Purpose: Research facility
Specialties: One of the first EGS experiments worldwide
Source of information: Sigfússon and Uihlein (2015b).

Sweden: Lund

Status: Concluded (start 2001)
Lithology: Precambrian igneous rocks (gneiss, amphibolite, metaba-
                            site, dolerite)
Wells: Exploration well: 3701.8 m
Reservoir: 85 °C, insufficient fluid production rate
Capacity: None
Purpose: Research facility
Specialties:

                            • Drilled into the Romeleåsen thrust fault zone
                            • Exploration well found too low temperature and fluid
                              production rate for a commercial use

Source of information: Rosberg and Erlström (2019).

Sweden: Malmö

Status: Ongoing (exploration phase)

Lithology: Precambrian igneous rocks: gneiss, amphibolite, metaba-
                            site, dolerite
Wells: Exploration well: 3133 m and 6000 m deep wells planned
Reservoir: 84.1 °C at 3133 m depth, geothermal gradient between 17.4
                            and 23.5 °C km^{-1}
Capacity: Not known
Purpose: Commercial plant
Specialties: District heating for the city of Malmö
Source of information: Rosberg and Erlström (2021).

Switzerland: Deep Heat Mining Project, Basel

Status: Abandoned (2005–2009)
Lithology: Igneous, Granite
Wells: Exploration: 2700 m; Basel 1: 5003 m
Reservoir: 200 °C, 70 kg s^{-1} (expected)
Capacity: 3 MW$_e$ and 20 MW$_{th}$ (planned)
Purpose: Commercial plant
Funding: CHF 28 Mio. from canton Basel, CHF 56 Mio. total project
                            costs
Specialties:

- Hydraulic stimulation below 4629 m depth
- Microseismic activity built up during first 6 days of
  fluid injection (magnitudes up to M2.6), therefore
  injection was stopped and the project had to be abandoned

Source of information: Sigfússon and Uihlein (2015b).

Switzerland: Haute-Sorne

Status: On-hold (planning phase)
Lithology: Granite, Gneiss
Wells: 4000–5000 m
Reservoir: >140 °C, >60 l s^{-1} (expected)
Capacity: max. 5 MW$_e$ (planned)
Purpose: Commercial plant
Funding: Supported by SF 90 Mio. from Switzerland
Specialties: Project delay due to legal process initiated by opponents,
                            which was won by the operator Geo-Energie Suisse. Yet,
                            local authorities withdrew concession in April 2020, future
                            of the project is uncertain

Source of information: Geo-Energie Suisse and Geo-Energie Jura (2019) and
                          Richter (2021).

UK: Eden Project, Cornwall

Status: Ongoing (development phase, drilling of the first well)
Lithology: Granite
Wells: Two 4500 m wells (planned)
Reservoir: Expected: 180–190 °C, 551 s^{-1}; hydraulic fracturing
Capacity: Planned: 4 MW$_e$
Purpose: Commercial plant
Funding: GBP 9.9 Mio. from the European Regional Development
                          Fund, GBP 1.4 Mio. from Cornwall Council, GBP 5.5 Mio.
                          from institutional investors
Specialties:

                          • Drilling into a fault system
                          • First phase: heating greenhouses and offices with a
                            coaxial circulation system in one well
                          • Second phase: second well and electricity plant

Source of information: Eden Geothermal (2021).

UK: United Downs Deep Geothermal Power Project, Cornwall

Status: Ongoing (construction phase)
Lithology: Granite
Wells: Production: 5275 m, injection: 2393 m
Reservoir: Expected: 190 °C, hydraulic fracturing planned
Capacity: Planned: 10 MW$_e$, 55 MW$_{th}$
Purpose: Commercial plant
Funding: Mixture of private and public funds
Specialties:

                          • Wells intersect the Porthtowan Fault Zone
                          • Binary power plant to be constructed

Source of information: GEL (2021).

UK: Rosemanowes

Status: Concluded (1984–1992)
Lithology: Granite
Wells: 2600 m

Reservoir: 79–100 °C, 4–25 l s^{-1}
Capacity: None
Purpose: Research facility
Specialties:

- One of the first EGS experiments worldwide
- Hydraulic fracturing, viscous gel stimulation, placement of proppants in joints
- Seismicity: max. M3.1

Source of information: Sigfússon and Uihlein (2015b).

EGS—Hot Sedimentary Aquifer

Germany: Bruchsal

Status: Ongoing
Lithology: Sedimentary: Middle Bunter Sandstone
Wells: 1900 and 2450 m
Reservoir: 120–130 °C, 24 l s^{-1}
Capacity: 5.5 MW$_{th}$, 0.55 MW$_e$
Purpose: Commercial plant
Funding: Public subsidies: €2.5 Mio. EU, €2.7 Mio. Germany
Specialties: €8.1 Mio. drilling costs
Source of information: Sigfússon and Uihlein (2015b).

Germany: Landau

Status: Ongoing
Lithology: Sedimentary: Muschelkalk
Wells: 3170–3300 m
Reservoir: 159 °C, 70–80 l s^{-1}, hydraulic stimulation for injector
Capacity: 3 MW$_e$, 3–6 MW$_{th}$
Purpose: Commercial Plant
Specialties:

- Seismic events occurred in 2009 (2.4–2.7 M)
- Heaving and horizontal movements of the ground in 2013 led to temporary shut down

Source of information: Sigfússon and Uihlein (2015b).

Germany: Insheim

Status:	Ongoing (start: 2007, production since 2012)
Lithology:	Sedimentary: Keuper, Perm, Bunter sandstone
Wells:	3600–3800 m
Reservoir:	165 °C, 50–80 $l\,s^{-1}$, hydraulic stimulation
Capacity:	4.8 MW_e; planned: 6–10 MW_{th}
Purpose:	Commercial plant
Specialties:	Seismicity: 2–2.4 M
Source of information:	Sigfússon and Uihlein (2015b).

Germany: Unterhaching

Status:	Ongoing (since 2004)
Lithology:	Limestone
Wells:	3350–3380 m
Reservoir:	123 °C, 150 $l\,s^{-1}$, chemical stimulation for increasing the natural flow rates
Capacity:	3.4 MW_e, 38 MW_{th} (max.)
Purpose:	Commercial plant
Specialties:	Kalina plant
Source of information:	Sigfússon and Uihlein (2015b).

Hydrothermal Systems

Iceland: Hellisheidi

Status:	Ongoing (since 2001)
Lithology:	Basalt
Wells:	47 production wells, 17 re-injection wells, around 2000 m max. depth
Reservoir:	Flash steam
Capacity:	303 MW_e, 133 MW_{th}, extension of thermal plant to 400 MW possible
Purpose:	Commercial plant
Funding:	Total costs USD 278 Mio.
Source of information:	Gunnlaugson (2012).

Italy: Lardarello-Travale Area (Lago, Molinetto, Gabbro, Travale)

Status:	Ongoing

Lithology: Carbonate rocks
Wells: More than 500 wells, 500–4000 m
Reservoir: Dry steam; Steam flow: 22.22–69.44 kg s^{-1};
Inlet steam temperature: 127–195 °C
Capacity: Around 8 to >40 MW
Purpose: Commercial plants
Specialties: First geothermal site (1904), dry steam plants
Source of information: DiPippo (2016) and Minissale (1991).

Netherlands: The Hague

Status: Ongoing
Lithology: Sandstone
Wells: 2200 m doublet
Reservoir: 78 °C, around 42 l s^{-1}
Capacity: 6 MW$_{th}$
Purpose: Commercial plant
Specialties: District heating
Source of information: IF Technology (2021) and Richter (2020)

References

Ádám L, Cladouhos TT (2016) Challenges of South Hungarian EGS demonstration project. European Geothermal Congress 2016, Strasboug, France, 19–24 Sept 2016

Adams BM, Kuehn TH, Bielicki JM, Randolph JB, Saar MO (2014) On the importance of the thermosiphon effect in CPG (CO$_2$ Plume Geothermal) power systems. Energy 69:409–418. https://doi.org/10.1016/j.energy.2014.03.032

Adams BM, Kuehn TH, Bielicki JM, Randolph JB, Saar MO (2015) A comparison of electric power output of CO$_2$ Plume Geothermal (CPG) and brine geothermal systems for varying reservoir conditions. Appl Energy 140:365–377. https://doi.org/10.1016/j.apenergy.2014.11.043

Adams BM, Vogler D, Kuehn TH, Bielicki JM, Garapati N, Saar MO (2021) Heat depletion in sedimentary basins and its effect on the design and electric power output of CO$_2$ Plume Geothermal (CPG) systems. Renew Energy 172:1393–1403. https://doi.org/10.1016/j.renene.2020.11.145

Adityatama D, Purba D, Muhammad F, Agustino V, Wiharlan H, Pasmeputra KK (2020) Slim hole drilling overview for geothermal exploration in indonesia: potential and challenges. In: Proceedings, 45th workshop on geothermal reservoir engineering, Standford University. https://pangea.stanford.edu/ERE/db/GeoConf/papers/SGW/2020/Adityatama.pdf. Accessed 25 Oct 2021

Al-Hameedi ATT, Alkinani HH, Dunn-Norman S, Al-Alwani MA, Alshammari AF, Alkhamis MM, Mutar RA, Al-Bazzaz WH (2020) Experimental investigation of environmentally friendly drilling fluid additives (mandarin peels powder) to substitute the conventional chemicals used in water-based drilling fluid. J Pet Explor Prod Technol 10(2):407–417. https://doi.org/10.1007/s13202-019-0725-7

Aposteanu A, Berre I, Bertani R, Clauser C, Jaudin F, Kujbus A, Sanner B, Urchueguia J (2014) Geothermal technology roadmap. Technical report, European technology platform on renewable heating and cooling. https://www.rhc-platform.org/content/uploads/2020/02/Geothermal_Roadmap-WEB.pdf. Accessed 25 Oct 2021

Ásmundsson R, Pezard P, Sanjuan B, Henninges J, Deltombe JL, Halladay N, Lebert F, Gadalia A, Millot R, Gibert B, Violay M, Reinsch T, Naisse JM, Massiot C, Azaïs P, Mainprice D, Karytsas C, Johnston C (2014) High temperature instruments and methods developed for supercritical geothermal reservoir characterisation and exploitation—The HiTI project. Geothermics 49:90–98. https://doi.org/10.1016/j.geothermics.2013.07.008

Atrens AD, Gurgenci H, Rudolph V (2009) CO_2 Thermosiphon for Competitive Geothermal Power Generation. Energy & Fuels 23(1):553–557. https://doi.org/10.1021/ef80/0601z

Atrens AD, Gurgenci H, Rudolph V (2010) Electricity generation using a carbon-dioxide thermosiphon. Geothermics 39(2):161–169. https://doi.org/10.1016/j.geothermics.2010.03.001

Azari V, Al Badi M, Vazquez O, Al-Kalbani M, Mackay E (2020) Scale treatment optimization in geothermal wells. In: SPE Europec featured at 82nd EAGE conference and exhibition, society of petroleum engineers, Amsterdam, The Netherlands. https://doi.org/10.2118/200565-MS

Baria R, Baumgärtner J, Gérard A, Jung R, Garnish J (1999) European HDR research programme at Soultz-sous-Forêts (France) 1987–1996. Geothermics 28(4):655–669. https://doi.org/10.1016/S0375-6505(99)00036-X

Baujard C, Genter A, Dalmais E, Maurer V, Hehn R, Rosillette R, Vidal J, Schmittbuhl J (2017a) Hydrothermal characterization of wells GRT-1 and GRT-2 in Rittershoffen, France: implications on the understanding of natural flow systems in the rhine graben. Geothermics 65:255–268. https://doi.org/10.1016/j.geothermics.2016.11.001

Baujard C, Hehn R, Geneter A, Teza D, Baumgärtner J, Guinot F, Martin A, Steinlechner S (2017b) Rate of penetration of geothermal wells: a key challenge in hard rocks. In: PROCEEDINGS, 42nd Workshop on geothermal reservoir engineering, Stanford University, Stanford, California, February, 13–15, 2017. https://pangea.stanford.edu/ERE/pdf/IGAstandard/SGW/2017/Baujard.pdf. Accessed 25 Oct 2021

Bello O, Holzmann J, Yaqoob T, Teodoriu C (2015) Application of artificial intelligence methods in drilling system design and operations: a review of the state of the art. J Artif Intell Soft Comput Res 5(2):121–139. https://doi.org/10.1515/jaiscr-2015-0024

Bertani R, Leray B, Wees JDV (eds) (2018) Vision for deep geothermal. european technology and innovation platform on deep geothermal (ETIP-DG). https://www.etip-dg.eu/front/wp-content/uploads/ETIP-DG_Vision_web.pdf. Accessed 4 Aug 2020

Bilić T, Raos S, Ilak P, Rajšl I, Pašičko R (2020) Assessment of geothermal fields in the South Pannonian basin system using a multi-criteria decision-making tool. Energies 13(5):1026. https://doi.org/10.3390/en13051026

Breede K, Dzebisashvili K, Liu X, Falcone G (2013) A systematic review of enhanced (or engineered) geothermal systems: past, present and future. Geoth Energy 1(1):4. https://doi.org/10.1186/2195-9706-1-4

Brown DW (2000) A hot dry rock geothermal energy concept utilizing supercritical CO_2 instead of water. In: Twenty-Fifth Workshop on geothermal reservoir engineering, Stanford, California. https://pangea.stanford.edu/ERE/pdf/IGAstandard/SGW/2000/Brown.pdf. Accessed 25 April 2021

Bruhn D, Manzella A, Vuataz F, Faulds J, Moeck I, Erbas K (2010) Exploration methods (chap 2). Geothermal energy systems. Wiley, New York, pp 37–112. https://doi.org/10.1002/9783527630479.ch2

Buonasorte G, Caravova R, Fiordelisi A, Ungarelli C (2013) MT as a tool for geothermal exploration: a case study from Southern Tuscany. European geothermal congress 2013, Pisa, Italy, 3–7 June 2013

Ciriaco AE, Zarrouk SJ, Zakeri G (2020) Geothermal resource and reserve assessment methodology: Overview, analysis and future directions. Renew Sust Energy Rev 119:109515. https://doi.org/10.1016/j.rser.2019.109515

Dalla Santa G, Galgaro A, Sassi R, Cultrera M, Scotton P, Mueller J, Bertermann D, Mendrinos D, Pasquali R, Perego R, Pera S, Di Sipio E, Cassiani G, De Carli M, Bernardi A (2020) An updated ground thermal properties database for GSHP applications. Geothermics 85:101758. https://doi.org/10.1016/j.geothermics.2019.101758

Deb P, Knapp D, Marquart G, Clauser C, Trumpy E (2020) Stochastic workflows for the evaluation of Enhanced Geothermal System (EGS) potential in geothermal greenfields with sparse data: The case study of Acoculco. Mexico Geoth 88:101879. https://doi.org/10.1016/j.geothermics.2020. 101879

Deb P, Salimzadeh S, Vogler D, Düber S, Clauser C, Settgast RR (2021) Verification of coupled hydraulic fracturing simulators using laboratory-scale experiments. Rock Mech Rock Eng. https:// doi.org/10.1007/s00603-021-02425-y

Deichmann N, Giardini D (2009) Earthquakes induced by the stimulation of an enhanced geothermal system below basel (Switzerland). Seismolog Res Lett 80(5):784–798. https://doi.org/10.1785/ gssrl.80.5.784

Dezayes C, IMAGE SP3 Team (2019) A new geothermal exploration workflow for deep sedimentary basins and basement. European geothermal congress 2019, Den Haag, The Netherlands, 11–14 June 2019

DiPippo R (2016) Larderello dry-steam power plants, Tuscany, Italy. In: DiPippo R (ed) Geothermal power plants (Fourth Edition), Butterworth-Heinemann, Oxford, pp 321–343. https://doi.org/10. 1016/B978-0-08-100879-9.00011-2

Doughty C, Dobson PF, Wall A, McLing T, Weiss C (2018) GeoVision analysis supporting task force report: exploration. Technical report LBNL-2001120, Lawrence Berkeley National Laboratory. https://escholarship.org/uc/item/4v7054cw. Accessed 11 Oct 2021

Dumas P, Antics M, Ungemach P (2013) Report on geothermal drilling. GEOELEC deliverable D3.3, European geothermal energy council (EGEC). http://www.geoelec.eu/wp-content/uploads/ 2011/09/D-3.3-GEOELEC-report-on-drilling.pdf

Eavor (2020) Eavor announces a commercial Eavor-Loop project to be built in Geretsried, Germany. https://www.eavor.com/press/eavor-announces-a-commercial-eavor-loop-project-to-be-built-in-geretsried-germany/. Accessed 8 Oct 2021

Eden Geothermal (2021) Eden Geothermal. https://www.edengeothermal.com/. Accessed 25 Oct 2021

Ekeinde EB, Okoro EE, Dosunmu A, Iyuke S (2019) Optimizing aqueous drilling mud system viscosity with green additives. J Pet Explor Prod Technol 9(1):315 318. https://doi.org/10.1007/ s13202-018-0481-0

ETIP-DG ETaIPoDG (2018) Strategic research and innovation agenda. Technical report, European technology and innovation platform on deep geothermal (ETIP-DG). http://www.etip-dg. eu/front/wp-content/uploads/AB_AC_ETIP-DG_SRA_v3.3_web.pdf. Accessed 25 Oct 2021

EU-FIRE (2016) EGS Hungary - The result of the pilot project http://egs-hungary.hu/projekt/pilot-projekt-eredmenye. Accessed 25 Oct 2021

Evans KF, Genter A, Sausse J (2005) Permeability creation and damage due to massive fluid injections into granite at 3.5 km at Soultz: 1. Borehole observations. J Geophys Res: Solid Earth 110(B4). https://doi.org/10.1029/2004JB003168

Eyerer S, Schifflechner C, Hofbauer S, Bauer W, Wieland C, Spliethoff H (2020) Combined heat and power from hydrothermal geothermal resources in Germany: an assessment of the potential. Renew Sust Energy Rev 120:109661. https://doi.org/10.1016/j.rser.2019.109661

Ezzat M, Vogler D, Saar MO, Adams BM (2021) Simulating plasma formation in pores under short electric pulses for plasma pulse geo drilling (PPGD). Energies 14(16):4717. https://doi.org/10. 3390/en14164717

Fleming MR, Adams BM, Kuehn TH, Bielicki JM, Saar MO (2020) Increased power generation due to exothermic water exsolution in CO_2 plume geothermal (CPG) power plants. Geothermics 88:101865. https://doi.org/10.1016/j.geothermics.2020.101865

Foley P, Skeehan K, Smith J, Mink R, Geohydro M (2016) Recovery Act. Direct confirmation of commercial geothermal resources in Colorado using remote sensing and on-site exploration, testing, and analysis. Technical report DOE-PAGOSA–0002828, 1238072, USDOE Office of energy efficiency and renewable energy (EERE), Geothermal Technologies Office. https://doi. org/10.2172/1238072

Garabetian T (2019) Report on Competitiveness of the geothermal industry. Technical report, European technology and innovation platform on deep geothermal (ETIP-DG). http://www.etip-dg. eu/front/wp-content/uploads/D4.6-Report-on-Competitiveness.pdf. Accessed 25 Oct 2021

Garapati N, Randolph JB, Saar MO (2015) Brine displacement by CO_2, energy extraction rates, and lifespan of a CO_2-limited CO_2-Plume Geothermal (CPG) system with a horizontal production well. Geothermics 55:182–194. https://doi.org/10.1016/j.geothermics.2015.02.005

Gaucher E, Schoenball M, Heidbach O, Zang A, Fokker P, Van Wees J, Kohl T (2015) Induced seismicity in geothermal reservoirs: a review of forecasting approaches. Renew Sustain Energy Rev 52:1473–1490. https://doi.org/10.1016/j.rser.2015.08.026

Gehringer M, Loksha V (2012) Geothermal handbook: planning and financing power generation. Technical report 002/12, World Bank, Washington, DC. https://openknowledge.worldbank.org/ handle/10986/23712. Accessed 19 Aug 2020

GEL (2021) United downs geothermal power project. https://geothermalengineering.co.uk/united-downs/. Accessed 25 Oct 2021

Genter A, Evans K, Cuenot N, Fritsch D, Sanjuan B (2010) Contribution of the exploration of deep crystalline fractured reservoir of Soultz to the knowledge of enhanced geothermal systems (EGS). Comptes Rendus Geosci 342(7):502–516. https://doi.org/10.1016/j.crte.2010.01.006

Gentier S, Rachez X, Dezayes C, Blaisonneau A, Genter A (2005) How to understand the effect of the hydraulic stimulation in terms of hydro-mechanical behavior at Soultz-sous-Forêts (France). GRC Trans 29:9. https://www.researchgate.net/publication/236622898_How_to_understand_ the_effect_of_the_hydraulic_stimulation_in_term_of_hydro-mechanical_behavior_at_Soultz-sous-Forets_France

Suisse G-E, Jura G-E (2019) Deep geothermal pilot project Haute-Sorne. https://www.geo-energie-jura.ch/le-projet/. Accessed 25 Oct 2021

GEOVEN (2021) GEOVEN Vendenheim. http://www.geoven.fr/. Accessed 25 Oct 2021

Gerard A, Kappelmeyer O (1991) European HDR Project at Soultz sous Forêts. Geoth Sci Technol 2 3(1-4):263–289. https://www.geothermal-library.org/index.php?mode=pubs&action=view& record=1009683. Accessed 23 Oct 2021

Gischig VS, Giardini D, Amann F, Hertrich M, Krietsch H, Loew S, Maurer H, Villiger L, Wiemer S, Bethmann F, Brixel B, Doetsch J, Doonechaly NG, Driesner T, Dutler N, Evans KF, Jalali M, Jordan D, Kittilä A, Ma X, Meier P, Nejati M, Obermann A, Plenkers K, Saar MO, Shakas A, Valley B (2020) Hydraulic stimulation and fluid circulation experiments in underground laboratories: stepping up the scale towards engineered geothermal systems. Geomech Energy Environ 24:100175. https://doi.org/10.1016/j.gete.2019.100175

Grant MA, Clearwater J, Quinao J, Bixley PF, Le Brun M (2013) Thermal stimulation of geothermal wells: a review of field data. In: PROCEEDINGS, Stanford University, Stanford, California, p 7. https://pangea.stanford.edu/ERE/pdf/IGAstandard/SGW/2013/Grant1.pdf. Accessed 20 Oct 2021

Griffiths L, Heap MJ, Baud P, Schmittbuhl J (2017) Quantification of microcrack characteristics and implications for stiffness and strength of granite. Int J Rock Mech Mining Sci 100:138–150. https://doi.org/10.1016/j.ijrmms.2017.10.013

Gunnlaugson E (2012) The Hellisheidi geothermal project - financial aspects of geothermal development. In: Short course on geothermal development and geothermal wells, United Nations University Geothermal Training Programme and LaGeo, Santa Tecla, El Salvador. https://orkustofnun. is/gogn/unu-gtp-sc/UNU-GTP-SC-14-12.pdf. Accessed 23 Oct 2021

Harvey C, Beardsmore G (eds) (2014) Best practice guide for geothermal exploration. IGA Service GmbH, Bochum. https://documents1.worldbank.org/curated/en/190071480069890732/ pdf/110532-Geothermal-Exploration-Best-Practices-2nd-Edition-FINAL.pdf. Accessed 19 Oct 2021

Heimlich C, Gourmelen N, Masson F, Schmittbuhl J, Kim SW, Azzola J (2015) Uplift around the geothermal power plant of Landau (Germany) as observed by InSAR monitoring. Geotherm Energy 3(1):2. https://doi.org/10.1186/s40517-014-0024-y

Herzberger P, Münch W, Kölbel T, Bruchmann U, Schlagermann P, Hötzl H, Wolf L, Rettenmaier D, Steger H, Zorn R, Seibt P (2010) The geothermal power plant Bruchsal. In: Proceedings, Bali, Indonesia, p 6. https://www.geothermal-energy.org/pdf/IGAstandard/WGC/2010/0619.pdf. Accessed 23 Oct 2021

Hettkamp T, Teza D, Baumgärtner J, Gandy T, Homeier G (2007) A multi-horizon approach for the exploration and exploitation of a fractured geothermal reservoir in Landau/Palatinate. In: First European geothermal review, Mainz, Germany, p 2

Hillers G, Vuorinen TAT, Uski MR, Kortström JT, Mäntyniemi PB, Tiira T, Malin PE, Saarno T (2020) The 2018 geothermal reservoir stimulation in Espoo/Helsinki, Southern Finland: seismic network anatomy and data features. Seismolog Res Lett 91(2A):770–786. https://doi.org/10.1785/0220190253

Hofmann H, Zimmermann G, Zang A, Min KB (2018) Cyclic soft stimulation (CSS): a new fluid injection protocol and traffic light system to mitigate seismic risks of hydraulic stimulation treatments. Geotherm Energy 6(1):27. https://doi.org/10.1186/s40517-018-0114-3

Hofmann H, Zimmermann G, Farkas M, Huenges E, Zang A, Leonhardt M, Kwiatek G, Martinez-Garzon P, Bohnhoff M, Min KB, Fokker P, Westaway R, Bethmann F, Meier P, Yoon KS, Choi JW, Lee TJ, Kim KY (2019) First field application of cyclic soft stimulation at the Pohang enhanced geothermal system site in Korea. Geophys J Int 217(2):926–949. https://doi.org/10.1093/gji/ggz058

Huenges E, Ledru P (2011) Geothermal energy systems: exploration, development, and utilization. Wiley, New York

IF Technology (2021) Deep geothermal for the Hague. https://www.iftechnology.com/project/deep-geothermal-for-the-hague/. Accessed 25 Oct 2021

IRENA and Aalborg University (2021) Integrating low-temperature renewables in district energy systems: guidelines for policy makers. International Renewable Energy Agency (IRENA), Aalborg University, Abu Dhabi, Copenhagen. https://www.irena.org/publications/2021/March/Integrating-low-temperature-renewables-in-district-energy-systems. Accessed 23 Oct 2021

Islam MR, Hossain ME (2021) Chapter 3 - Advances in directional drilling. In: Islam MR, Hossain ME (eds) Drilling engineering, sustainable oil and gas development series. Gulf Professional Publishing, Elsevier, pp 179–316. https://doi.org/10.1016/B978-0-12-820193-0.00003-4

Jamali S, Wittig V, Börner J, Bracke R, Ostendorf A (2019) Application of high powered Laser Technology to alter hard rock properties towards lower strength materials for more efficient drilling, mining, and Geothermal Energy production. Geomech Energy Environ 20:100112. https://doi.org/10.1016/j.gete.2019.01.001

Kolditz O, Bauer S, Bilke L, Böttcher N, Delfs JO, Fischer T, Görke UJ, Kalbacher T, Kosakowski G, McDermott CI, Park CH, Radu F, Rink K, Shao H, Shao HB, Sun F, Sun YY, Singh AK, Taron J, Walther M, Wang W, Watanabe N, Wu Y, Xie M, Xu W, Zehner B (2012) OpenGeoSys: An open-source initiative for numerical simulation of thermo-hydro-mechanical/chemical (THM/C) processes in porous media. Environ Earth Sci 67(2):589–599. https://doi.org/10.1007/s12665-012-1546-x

Kratt C (2011) Hyperspectral, shallow temperature, and gravity surveys: a roundup of recent exploration activity at Silver Peak, Alum, and Columbus Salt Marsh, Esmeralda County, Nevada. Geotherm Resour Council Trans 35:853–860

Kujbus A, van Gelder G, Urchueguia JF, Pockelé L, Guglielmetti L, Bloemendal M, Blum P, Pasquali R, Bonduà S (2020) Strategic research innovation agenda for geothermal technologies. Technical report, Geothermal panel of European technology platform on renewable heating and cooling. www.rhc-platform.org. Accessed 21 Oct 2021

Kukkonen IT, Pentti M (2021) St1 Deep heat project: geothermal energy to the district heating network in Espoo. IOP Conf Ser: Earth Environ Sci 703(1):012035. https://doi.org/10.1088/1755-1315/703/1/012035

Kullick J, Hackl C (2017) Dynamic modeling and simulation of deep geothermal electric submersible pumping systems. Energies 10(10):1659. https://doi.org/10.3390/en10101659

Kumari WGP, Ranjith PG (2019) Sustainable development of enhanced geothermal systems based on geotechnical research – a review. Earth-Sci Rev 199:102955. https://doi.org/10.1016/j.earscirev.2019.102955

Leonhardt M, Kwiatek G, Martínez-Garzón P, Bohnhoff M, Saarno T, Heikkinen P, Dresen G (2021) Seismicity during and after stimulation of a 6.1 km deep enhanced geothermal system in Helsinki, Finland. Solid Earth 12(3):581–594. https://doi.org/10.5194/se-12-581-2021

Li T, Shiozawa S, McClure MW (2016) Thermal breakthrough calculations to optimize design of a multiple-stage enhanced geothermal system. Geothermics 64:455–465. https://doi.org/10.1016/j.geothermics.2016.06.015

Lu SM (2018) A global review of enhanced geothermal system (EGS). Renew Sustain Energy Rev 81:2902–2921. https://doi.org/10.1016/j.rser.2017.06.097

Mackenzie K, Ussher G, Libbey R, Quinlivan P, Dacanay J, Bogie I (2017) Use of deep slim-hole drilling for geothermal exploration. In: Proceedings, 5th Indonesia international geothermal convention & exhibition (IIGCE) 2017, 2–4 August, Jakarta, Indonesia

Macpherson JD, de Wardt JP, Florence F, Chapman C, Zamora M, Laing M, Iversen F (2013) Drilling-systems automation: current state, initiatives, and potential impact. SPE Drilling & Completion 28(04):296–308. https://doi.org/10.2118/166263-PA

Manente G, Lazzaretto A, Bardi A, Paci M (2019) Geothermal power plant layouts with water absorption and reinjection of H_2S and CO_2 in fields with a high content of non-condensable gases. Geothermics 78:70–84. https://doi.org/10.1016/j.geothermics.2018.11.008

Meixner J, Schill E, Gaucher E, Kohl T (2014) Inferring the in situ stress regime in deep sediments: an example from the Bruchsal geothermal site. Geotherm Energy 2(1):7. https://doi.org/10.1186/s40517-014-0007-z

Minissale A (1991) The Larderello geothermal field: a review - ScienceDirect. Earth-Sci Rev 31(2). https://doi.org/10.1016/0012-8252(91)90018-B

Mouchot J, Genter A, Cuenot N, Scheiber J, Seibel O, Bosia C, Ravier G (2018) First year of operation from EGS geothermal plants in Alsace, France: scaling issues. In: Proceedings, 43rd workshop on geothermal reservoir engineering, Stanford University, Stanford, California, February 12–14, 2018 SGP-TR-213

NASA (2021) Technology readiness level. https://www.nasa.gov/directorates/heo/scan/engineering/technology/technology_readiness_level. Accessed 26 Oct 2021

Niknam PH, Talluri L, Fiaschi D, Manfrida G (2020) Gas purification process in a geothermal power plant with total reinjection designed for the Larderello area. Geothermics 88:101882. https://doi.org/10.1016/j.geothermics.2020.101882

NN (2021) Plasma Pulse Geo Drilling (PPGD). https://geg.ethz.ch/project-plasma_drilling/. Accessed 25 Oct 2021

Nogara J, Zarrouk SJ (2018) Corrosion in geothermal environment: part 1: fluids and their impact. Renew Sustain Energy Rev 82:1333–1346. https://doi.org/10.1016/j.rser.2017.06.098

Pine RJ, Batchelor AS (1984) Downward migration of shearing in jointed rock during hydraulic injections. Int J Rock Mech Mining Sci & Geomech Abstr 21(5):249–263. https://doi.org/10.1016/0148-9062(84)92681-0

Planès T, Obermann A, Antunes V, Lupi M (2020) Ambient-noise tomography of the Greater Geneva Basin in a geothermal exploration context. Geophys J Int 220(1):370–383. https://doi.org/10.1093/gji/ggz457

Polsky Y, Jr LC, Finger J, Huh M, Knudsen S, Chip AJ, Raymond D, Swanson R (2008) Enhanced geothermal systems (EGS) Well construction technology evaluation report. Sandia Report SAND2008-7866, Sandia National Laboratories, Albuquerque, New Mexico and Livermore, California. https://www1.eere.energy.gov/geothermal/pdfs/egs_well_contruction.pdf. Accessed 25 Oct 2021

Portier ES, Vuataz FD, Evans K, Valley B, Häring M, Hopkirk RJ, Baujard C, Kohl T, Mégel T (2009a) Studies and support for the EGS reservoirs at Soultz-sous-Forêts. Final report p 112

Portier S, André L, Vuataz FD (2007) Review on chemical stimulation techniques in oil industry and applications to geothermal systems. Technical Report 1, Deep Heat Mining Association, Switzerland. http://citeseerx.ist.psu.edu/viewdoc/download?doi=10.1.1.588.2759&rep=rep1&type=pdf

Portier S, Vuataz FD, Nami P, Sanjuan B, Gérard A (2009) Chemical stimulation techniques for geothermal wells: experiments on the three-well EGS system at Soultz-sous-Forêts. France. Geothermics 38(4):349–359. https://doi.org/10.1016/j.geothermics.2009.07.001

Pruess K (2006) Enhanced geothermal systems (EGS) using CO_2 as working fluid—a novel approach for generating renewable energy with simultaneous sequestration of carbon. Geothermics 35(4):351–367. https://doi.org/10.1016/j.geothermics.2006.08.002

Pruess K (2008) On production behavior of enhanced geothermal systems with CO_2 as working fluid. Energy Convers Manag 49(6):1446–1454. https://doi.org/10.1016/j.enconman.2007.12.029

Rajšl I, Raos S, Bilić T (2019) Open access decision support tool for optimal usage of geothermal energy. MEET Deliverable D7.1, The MEET Consortium. https://www.meet-h2020.com/wp-content/uploads/2020/07/MEET_Deliverable_D7.1.pdf. Accessed 24 Oct 2021

Randolph JB, Saar MO (2011) Coupling carbon dioxide sequestration with geothermal energy capture in naturally permeable, porous geologic formations: implications for CO_2 sequestration. Energy Proc 4:2206–2213. https://doi.org/10.1016/j.egypro.2011.02.108

Raos S, Ilak P, Rajšl I, Bilić T, Trullenque G (2019) Multiple-criteria decision-making for assessing the enhanced geothermal systems. Energies 12(9):1597. https://doi.org/10.3390/en12091597

Riches S, Johnston C (2015) Electronics design, assembly and reliability for high temperature applications. In: 2015 IEEE International Symposium on Circuits and Systems (ISCAS), IEEE, Lisbon, Portugal, pp 1158–1161. https://doi.org/10.1109/ISCAS.2015.7168844

Richter A (2020) After challenging start, The Hague to tap into geothermal energy for heating. https://www.thinkgeoenergy.com/after-challenging-start-the-hague-to-tap-into-geothermal-energy-for-heating/. Accessed 25 Oct 2021

Richter A (2021) Government silence on decision for or against Haute-Sorne geothermal project, Switzerland. https://www.thinkgeoenergy.com/government-silence-on-decision-for-or-against-haute-sorne-geothermal-project-switzerland/. Accessed 25 Oct 2021

Richter M (2017) Summary of new drilling technologies. Technical report, International Energy Agency - Geothermal (IEA Geothermal). http://iea-gia.org/wp-content/uploads/2014/10/IEA-Geothermal-Drilling-Technologies.pdf. Accessed 25 Oct 2021

Rosberg JE, Erlström M (2019) Evaluation of the Lund deep geothermal exploration project in the Romeleåsen Fault Zone, South Sweden: a case study. Geotherm Energy 7(1). https://doi.org/10.1186/s40517-019-0126-7

Rosberg JE, Erlström M (2021) Evaluation of deep geothermal exploration drillings in the crystalline basement of the Fennoscandian Shield border zone in South Sweden. Geotherm Energy 9(1):20. https://doi.org/10.1186/s40517-021-00203-1

Rossi E, Jamali S, Saar MO, Rudolf von Rohr P (2020a) Field test of a combined thermo-mechanical drilling technology. Mode I: thermal spallation drilling. J Pet Sci Eng 190:107005. https://doi.org/10.1016/j.petrol.2020.107005

Rossi E, Jamali S, Schwarz D, Saar MO, Rudolf von Rohr P (2020b) Field test of a combined thermo-mechanical drilling technology. Mode II: flame-assisted rotary drilling. J Pet Sci Eng 190:106880. https://doi.org/10.1016/j.petrol.2019.106880

Rossi E, Jamali S, Wittig V, Saar MO, Rudolf von Rohr P (2020c) A combined thermo-mechanical drilling technology for deep geothermal and hard rock reservoirs. Geothermics 85:101771. https://doi.org/10.1016/j.geothermics.2019.101771

Rossi E, Saar MO, Rudolf von Rohr P (2020d) The influence of thermal treatment on rock–bit interaction: a study of a combined Thermo–mechanical Drilling (CTMD) concept. Geotherm Energy 8(1):16. https://doi.org/10.1186/s40517-020-00171-y

Saadat A, Frick S, Kranz S, Regenspurg S (2010) Energetic use of EGS reservoirs (chap 6). In: Geothermal energy systems. Wiley, New York, pp 303–372. https://doi.org/10.1002/9783527630479.ch6

Schindler M, Baumgärtner J, Gandy T, Hauffe P, Hettkamp T, Menzel H, Penzkofer P, Teza D, Tischner T, Wahl G (2010) Successful hydraulic stimulation techniques for electric power production in the Upper Rhine Graben, Central Europe. In: Proceedings, Bali, Indonesia, vol 2010, p 7. https://www.geothermal-energy.org/pdf/IGAstandard/WGC/2010/3163.pdf. Accessed 24 Oct 2021

Schmittbuhl J, Lambotte S, Lengliné O, Grunberg M, Jund H, Vergne J, Cornet F, Doubre C, Masson F (2021) Induced and triggered seismicity below the city of Strasbourg, France from November 2019 to January 2021. Comptes Rendus Géoscience Online first. https://doi.org/10.5802/crgeos.71

Seidler R, Padalkina K, Bücker HM, Ebigbo A, Herty M, Marquart G, Niederau J (2016) Optimal experimental design for reservoir property estimates in geothermal exploration. Comput Geosci 20(2):375–383. https://doi.org/10.1007/s10596-016-9565-4

Shook GM, Suzuki A (2017) Use of tracers and temperature to estimate fracture surface area for EGS reservoirs. Geothermics 67:40–47. https://doi.org/10.1016/j.geothermics.2016.12.006

Sigfússon B, Uihlein A (2015a) 2014 JRC geothermal energy status report - technology, market and economic aspects of geothermal energy in Europe. EUR - Scientific and Technical Research Reports, Institute for Energy and Transport (Joint Research Centre), Publications Office of the European Union, 10.2790/460251

Sigfússon B, Uihlein A (2015b) 2015 JRC geothermal energy status report - technology, market and economic aspects of geothermal energy in Europe. Technical report, Institute for Energy and Transport (Joint Research Centre), Publications Office of the European Union. https://doi.org/10.2790/959587

Sperber A, Moeck I, Brandt W (2010) Drilling into geothermal reservoirs (chap 3). In: Geothermal energy systems. Wiley, New York, pp 113–171. https://doi.org/10.1002/9783527630479.ch3

Spichak V, Manzella A (2009) Electromagnetic sounding of geothermal zones. J Appl Geophys 68(4):459–478. https://doi.org/10.1016/j.jappgeo.2008.05.007

St1 (2021) News about geothermal heat. http://st1.com/geothermal-heat. Accessed 25 Oct 2021

Stober I, Bucher K (2021a) Enhanced-Geothermal-Systems (EGS), Hot-Dry-Rock Systems (HDR), Deep-Heat-Mining (DHM). In: Stober I, Bucher K (eds) Geothermal energy: from theoretical models to exploration and development. Springer International Publishing, Cham, pp 205–225. https://doi.org/10.1007/978-3-030-71685-1_9

Stober I, Bucher K (2021b) Geophysical methods, exploration and analysis. In: Stober I, Bucher K (eds) Geothermal energy: from theoretical models to exploration and development. Springer International Publishing, Cham, pp 311–325. https://doi.org/10.1007/978-3-030-71685-1_13

Strack KM, Allegar N, Yu G, Tulinius H, Adam L, Gunnarsson Á, He L, He Z (2008) Mapping geothermal reservoirs using broadband 2-D MT and gravity data. In: Conference Proceedings, 70th EAGE conference and exhibition - workshops and fieldtrips, Rome, Italy, 9–12 June 2008. https://doi.org/10.3997/2214-4609.201404995

Sutra E, Spada M, Burgherr P (2017) Chemicals usage in stimulation processes for shale gas and deep geothermal systems: a comprehensive review and comparison. Renew Sustain Energy Rev 77:1–11. https://doi.org/10.1016/j.rser.2017.03.108

Tao J, Wu Y, Elsworth D, Li P, Hao Y (2019) Coupled thermo-hydro-mechanical-chemical modeling of permeability evolution in a CO_2-circulated geothermal reservoir. Geofluids 2019:1–15. https://doi.org/10.1155/2019/5210730

Teodoriu C, Cheuffa C (2011) A comprehensive review of past and present drilling methods with application to deep geothermal environment. In: Proceedings, Thirty-Sixth workshop on geothermal reservoir engineering, Stanford University, Stanford, California, January 31–February 2, 2011 SGP-TR-191. https://pangea.stanford.edu/ERE/pdf/IGAstandard/SGW/2011/teodoriu.pdf

Thorbjornsson I, Kaldal G, Gunnarsson B, Ragnarsson A (2017) A new approach to mitigate casing failures in high-temperature geothermal wells. GRC Transactions 41

Tischner T, Evers H, Hauswirth H, Jatho R, Kosinowski M (2010) New concepts for extracting geothermal energy from one well: the GeneSys-project. In: Proceedings world geothermal congress 2010, Bali, Indonesia, p 5. https://www.geothermal-energy.org/pdf/IGAstandard/WGC/2010/2272.pdf. Accessed 25 Oct 2021

Vallier B, Magnenet V, Schmittbuhl J, Fond C (2019) Large scale hydro-thermal circulation in the deep geothermal reservoir of Soultz-sous-Forêts (France). Geothermics 78:154–169. https://doi.org/10.1016/j.geothermics.2018.12.002

Verma MP, Torres-Encarnacion JA (2018) GeoSteam.Net: steam transport simulator for geothermal pipeline network. In: Proceedings, 43rd Workshop on geothermal reservoir engineering, Stanford University, Stanford, California, February 12–14, 2018 p 5. https://pangea.stanford.edu/ERE/pdf/IGAstandard/SGW/2018/Verma1.pdf. Accessed 24 Oct 2021

Vidal J, Genter A (2018) Overview of naturally permeable fractured reservoirs in the central and southern Upper Rhine Graben: insights from geothermal wells. Geothermics 74:57–73. https://doi.org/10.1016/j.geothermics.2018.02.003

Vidal J, Genter A, Schmittbuhl J (2016) Pre- and post-stimulation characterization of geothermal well GRT-1, Rittershoffen, France: insights from acoustic image logs of hard fractured rock. Geophys J Int 206(2):845–860. https://doi.org/10.1093/gji/ggw181

Vogt C, Marquart G, Kosack C, Wolf A, Clauser C (2012) Estimating the permeability distribution and its uncertainty at the EGS demonstration reservoir Soultz-sous-Forêts using the ensemble Kalman filter. Water Resour Res 48(8). https://doi.org/10.1029/2011WR011673

Vogt C, Iwanowski-Strahser K, Marquart G, Arnold J, Mottaghy D, Pechnig R, Gnjezda D, Clauser C (2013) Modeling contribution to risk assessment of thermal production power for geothermal reservoirs. Renew Energy 53:230–241. https://doi.org/10.1016/j.renene.2012.11.026

van Wees JD, Hopman J, Hersir G, Flóvenz O, Dezayes C, Manzella A, Bruhn D, Sippel J, Liotta D (2017) IMAGE - beyond standardized workflow for geothermal exploration. IMAGE Deliverable D2.05, The IMAGE Consortium. http://www.image-fp7.fr/reference-documents/Pages/default.aspx. Accessed 28 Oct 2021

Wertich V, Tiewsoh L, Tiess G (2018) Economic feasibility assessment methodology. Deliverable D5.2, CHPM2030 project, University of Miskolc, Hungary. https://www.chpm2030.eu/wp-content/uploads/2018/08/CHPM2030_D5.2.pdf. Accessed 26 Oct 2021

Wittig V, Brackc R, Hyun-Ick Y (2015) Hydraulic DTH fluid/mud hammers with recirculation capabilities to improve ROP and hole cleaning for deep, hard rock geothermal drilling. In: Proceedings world geothermal congress 2015, Melbourne, Australia, p 9

Wu Y, Li P (2020) The potential of coupled carbon storage and geothermal extraction in a CO_2-enhanced geothermal system: a review. Geotherm Energy 8(19). https://doi.org/10.1186/s40517-020-00173-w

Yuan W, Chen Z, Grasby SE, Little E (2021) Closed-loop geothermal energy recovery from deep high enthalpy systems. Renew Energy 177:976–991. https://doi.org/10.1016/j.renene.2021.06.028

Zarrouk SJ, McLean K (2019) Chapter 2 - Geothermal systems. In: Zarrouk SJ, McLean K (eds) Geothermal well test analysis. Academic, Cambridge, pp 13–38. https://doi.org/10.1016/B978-0-12-814946-1.00002-5

Zhang Y, Zhao GF (2020) A global review of deep geothermal energy exploration: from a view of rock mechanics and engineering. Geomech Geophys Geo-Energy Geo-Resour 6(1):4. https://doi.org/10.1007/s40948-019-00126-z

Chapter 4
Risks and Barriers

Abstract As every technical application, the exploitation of deep geothermal energy is associated with technical and non-technical risks, which we introduce together with mitigation strategies in this chapter. Major risks for geothermal projects are the high upfront costs, mainly for drilling, and the inevitable uncertainty of the subsurface conditions that may result in low heat extraction rates. In addition, lack of social acceptance and induced seismicity constitute risks that can lead to the termination of geothermal projects. Identifying and mitigating risks hence remain a key task for each phase of geothermal project development. In order to unify and simplify the process of risk identification and the definition of mitigation strategies, ongoing research focuses on the development of a specific risk register for deep geothermal projects. Reservoir stimulation is of paramount importance, especially for EGS projects, but at the same time the associated induced seismicity represents a major risk. Developing research therefore focuses on improving the process understanding of reservoir stimulation and developing improved stimulation techniques. Together with new monitoring strategies, this research aims to prevent strong seismic events caused by reservoir stimulation. Environmental impacts of geothermal heat use from deep reservoirs can be assessed by life-cycle assessment. One main source of emissions is, for example, the energy consumption of the drilling process. Recent developments towards a simplified life-cycle assessment tool will allow for a quicker, yet scientifically robust, evaluation of the environmental performance of deep geothermal energy plants.

Keywords Induced seismicity · Social acceptance · Life-cycle assessment · Greenhouse gas emissions · Uncertainty

4.1 Introduction

Several abandoned geothermal projects in Appendix 2 already show various potential risks of geothermal energy which may ultimately lead to the termination of projects. The risks can be of technical or non-technical nature. International standards such as the ISO 31000 (International Organization for Standardization 2018) provide guidance for managing project risks. The essential first step is risk identification.

© The Author(s), under exclusive license to Springer Nature Switzerland AG 2022   75
J. Fink et al., *State of the Art in Deep Geothermal Energy in Europe*,
SpringerBriefs in Earth System Sciences,
https://doi.org/10.1007/978-3-030-96870-0_4

Therefore, the research project Geo-Risk (see Appendix 1) is implementing a **risk register** for deep geothermal projects (Le Guénan et al. 2020). In the context of deep geothermal energy, they define risks as "events (or sources of uncertainty) preventing the proper development of the project considering the economic, technical, environmental and social points of view" (Le Guénan et al. 2020, p. 4). Thereby, every risk has a different likelihood and severity, which have to be analyzed specifically for each deep geothermal project. In the following, we give an overview of some risks that occurred in previous deep geothermal projects and discuss possible **mitigation strategies**. We refer to the Georisk Tool at https://www.georisk-project.eu/georisk-tool/ for a comprehensive list and explanations of the risks and mitigation strategies related to deep geothermal project development.

4.2 Non-technical Risks

Every project development stage (see Fig. 3.1) involves risks that can lead to the termination of the project. The overriding risk is therefore the **economical risk** borne by the investors. Special financing and insurance schemes have the goal of mitigating this risk and public money or subsidies are a way of distributing the economical risk (Sanyal et al. 2016). For example, the Geo-Risk project aims at developing such insurance or financial risk mitigation schemes.

Geothermal project planning involves uncertainties in forecasting heating and cooling demands over the course of a geothermal project's 30–50-year lifespan. A robust demand site forecast is required for the business plan. Therefore, a public or political energy transition plan should consider the dependencies between investments in geothermal district heating projects and renovation of buildings. The efficiency increase caused by the latter leads to changes on the heating and cooling demand site, which in turn affects the business plan of geothermal heating projects. This is addressed by the EU that required all member countries to provide their building renovation strategy for 2050 (IRENA 2021). A renovation strategy is part of the Green New Deal, because most of the buildings in the EU are older than 50 years and buildings are responsible for 40% of energy consumption (European Commission 2020). Long-term strategies provided by member states enable investors and project planners to forecast heating and cooling demand more solidly and reliably, reducing risk on the demand site.

Another non-technical risk and potential barrier that exists throughout the project life cycle is **social acceptance**. The public's perception of geothermal is often less positive than that of other renewable energies, such as solar or wind (Pellizzone et al. 2015). European studies on social acceptance of geothermal projects are, for example, Leucht et al. (2010), Reith et al. (2013), Pellizzone et al. (2015), and the comprehensive book "Geothermal Energy and Society" (Manzella et al. 2019). They recommend to mitigate risk by informing and involving the public throughout the project planning and development. The project developers should design concepts and campaigns for public communication alongside with the technical planning of the project. Transparency and knowledge sharing are essential for gaining public acceptance of a deep geothermal project. This holds especially for EGS projects, because the public is aware of and might be concerned about the risk of induced seismicity (more details in Sect. 4.3). For instance, the development of the EGS project in Haute-Sorne, Switzerland, had been delayed by a legal process initiated by opponents of the technology. Although the operating company Geo-Energie Suisse won the lawsuit, local authorities withdrew the concession in April 2020, leaving the project's future uncertain (see Appendix 2).

As demonstrated by the Swiss example, the **regulatory framework** poses another risk to geothermal projects. Missing or withdrawn concessions result in project delay at least, or even in its termination. The non-uniformity of regulatory frameworks across Europe is a barrier for Europe-wide project development and holds back both companies and investors (Angelino 2017).

4.3 Technical Risks

In addition to the potential risks of non-technical nature, geothermal projects face various technical risks. During the early project stage, a deep geothermal project faces **exploration risks** caused by geological uncertainty. The exploration risks can be mitigated or reduced by a proper exploration campaign (see Sect. 3.2) resulting in a conceptual reservoir model and subsequent numerical reservoir modeling. Drilling exploration wells is the only way to significantly reduce the exploration risk (see Fig. 4.1). Especially for deep geothermal projects in crystalline basement rocks, a lot of information is only available after drilling into the reservoir. Lack of information from the exploration phase is also one of the highest risk sources for reservoir stimulation, since important factors such as the fracturing or physical and chemical parameters are unknown. Yet, already the temperature found at depth might be lower than expected before explorational drilling, as, for example, in Lund, Sweden (Rosberg and Erlström 2019, see Appendix 2). To a certain extent, there are technical solutions for mitigating the risk of discovering unfavorable reservoir conditions, provided the economic feasibility is given. One solution can be to drill deeper until the desired temperature is produced. Other solutions are, for example, to use heat pumps or innovative power plants for increasing the capacity at lower temperature. If the flow rate is too low, stimulation methods can be applied.

All technologies applied for accessing and developing a geothermal reservoir bear specific technical risks. Technical problems during drilling comprise, for instance, wellbore instability, stuck pipe, or fluid losses. Certain risks can be mitigated by choosing the appropriate drilling technology and method for the specific geological situation. Furthermore, accurate surface exploration before drilling helps to reduce technical risks, and monitoring while drilling (MWD) assists the drilling and decision-making process (Zarrouk and McLean 2019).

Many technical risks during reservoir development and operation are related to the site-specific subsurface characteristics. For example, problems with the technical

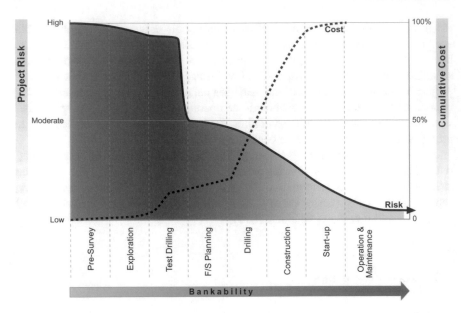

Fig. 4.1 Risk of a geothermal project versus the cumulative investment cost. Modified from Gehringer and Loksha (2012), used under CC BY 3.0 IGO, https://creativecommons.org/licenses/by/3.0/igo/

equipment and materials can occur due to high temperature or chemically aggressive environments.

A major technical risk for EGS projects is **induced seismicity**, which can be caused by reservoir stimulation (see also Sects. 1.4 and 3.4.1). Seismic activities triggered by geothermal exploitation have been observed around geothermal sites worldwide over the last 40 years (e.g. Zang et al. 2014). The issue of anthropogenic seismicity is also known from other subsurface energy operations, such as oil and gas or mining (Porter et al. 2019), as depicted in Fig. 4.2. However, one should pay special attention to the magnitude of the events (see Sect. 1.4, Table 1.1). Induced seismicity may be entirely of very low magnitudes (microseismicity), or may be a short-lived transient phenomenon. For example, the stimulation of the 6 km deep wells in Espoo (Finland) in 2018 induced more than 50000 micro-earthquakes with magnitudes below M1.9, most events had magnitudes around M0.0 (Leonhardt et al. 2021; Kukkonen and Pentti 2021). In contrast, the highest intensity of a seismic event triggered by EGS stimulation of 5.5 on Richter scale was measured in Pohang, Korea, in 2017 (Grigoli et al. 2018; Ellsworth et al. 2019; Kim et al. 2020). This magnitude indicates a seismic event potentially felt by everyone, which causes significant damage to infrastructure. It highlights that induced seismicity is a major issue, which needs to be considered. Several mechanisms can induce seismic activity such as the increased fluid pressure due to hydraulic stimulation which facilitates seismic slip in the presence of an unbalanced stress field (Cuenot et al. 2011) or thermo-elastic strain through thermal stimulation (Candela et al. 2018).

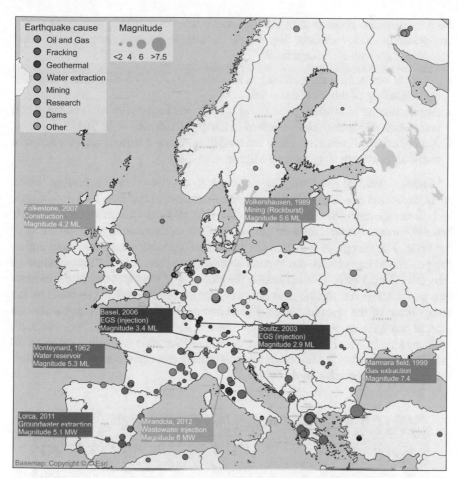

Fig. 4.2 Overview of induced and triggered seismic events in Europe related to industrial activities with recorded magnitude ≥ 1.5 (after an idea by Porter et al. 2019). Data products for this map were accessed through The Human-Induced Earthquake Database (HiQuake, Wilson et al. 2017; Foulger et al. 2018, http://inducedearthquakes.org, last accessed October 27, 2021)

Induced seismicity is one main reason for the termination of EGS projects, especially in densely populated urban areas. Seismic events with a magnitude greater than 2 have raised concerns among residents about the damage caused by individual events as well as their cumulative effects. In Basel (Switzerland), for example, microseismic events in 2006 lead to the termination of the project (e.g. Mukuhira et al. 2013; Edwards et al. 2015). In 2020, the GEOVEN project in Strasbourg (France) was terminated after triggered earthquakes of magnitudes 3.0 and 3.6 ML, which were above the red-light limit (M2.0) of the applied traffic-light system (Schmittbuhl et al. 2021). Eventually, a reduction of the risk of induced seismicity and successful demonstration projects with low seismicity would contribute to a better public acceptance of EGS. However, they should include the following measures to avoid seismic events with high magnitudes:

- A proper exploration and planning as described in Chap. 3 resulting in a thorough understanding of the local stress field.
- Application of innovative and careful stimulation techniques such as the cyclic soft stimulation techniques described in Chap. 3 (Hofmann et al. 2019).
- A monitoring (Hillers et al. 2020) and modeling strategy, for example, the adaptive traffic-light system suggested by Mignan et al. (2017).

Another technical risk during the drilling and operation phases is **scaling**, which is the deposition of minerals in the well pipes, the pump, pipelines, heat exchangers, and filters (Sperber et al. 2010, Fig. 4.3). The minerals precipitate from the reservoir brine due to redox reactions with the metal of the well or due to over-saturation of the brine. The change of pressure, pH value, flow dynamics or temperature during the production can change the mineral solubility and result in the precipitation of minerals (Corsi 1986). This may cause the clogging of the pipes and of the production pump. However, scaling does not occur for every reservoir but depends on the composition of the reservoir brine (Sperber et al. 2010). For example, calcite and siderite precipitation is observed for carbonated fluids; dissolved sulfites can react with any kind of metal (such as Pb, Cu, or Fe) used for tubing, forming the minerals galena and sphalerite. Additionally, massive scaling can occur if oxygen (O_2) enters ferrous solutions forming ferric oxides.

Fig. 4.3 Precipitation of minerals from a geothermal brine within a pipe. This so-called scaling successively reduces flow rates up to the clogging of the pipe. The picture was taken in the Geothermal Museum in Lardarello, Italy. Reproduced from Pluymakers (2019) with permission from Anne Pluymakers

There are several methods to avoid or mitigate scaling (Sperber et al. 2010). A major factor for mitigating scaling effects is a good understanding of the fluid chemistry in the reservoir. Especially thermo-hydro-chemical (THC) simulations can help to reduce scaling by optimizing operation or materials (Garcia et al. 2005; Xu and Pruess 2001).

4.4 Environmental Risks

Last but not least, **environmental risks** associated with geothermal projects are considered often in conjunction with reservoir brine and other aspects of operations. These risks comprise, for example:

- Disturbance at the surface due to land use, noise, dust, or smell.
- Degassing of produced reservoir fluids.
- Disturbance of groundwater aquifers caused by well integrity failures or poor cementing.
- Energy and water consumption during drilling and surface operations.

The life-cycle assessment (LCA) methodology provides a tool for mitigating risks of environmental pollution and associated concerns by the public. Currently, the research project GEOENVI (see Appendix 1) is developing a simplified LCA tool for assessing the environmental impacts and benefits of deep geothermal energy plants within one day. They published LCA guidelines for geothermal installation recently (Blanc et al. 2020). In addition, they built a database of environmental risks and their consequences and link them to the different project development stages. We refer to the GEOENVI database that is available as an online tool at https:// geoenvi.brgm.fr/environmental_aspects for detailed description of the environmental risks and mitigation strategies. The main goal of most regulatory frameworks for geothermal projects is to prevent or minimize environmental impact. To this end, they require careful planning and continuous monitoring.

Greenhouse gas (GHG) emissions are a particular environmental impact that is often the focus of renewable energy assessments. GHG emissions are usually divided into emissions from the construction phase (indirect sources) and from the operation phase (direct sources). Generally, the indirect emissions are higher than the direct emissions for all renewable energy sources, whereas for fossil energy sources, it is vice versa. The LCA summarizes all emissions from the beginning to the end of a power plant. The ratio of produced emissions and produced energy in g CO_2 eq/kWh is typically used to compare different energy sources and plants. Ideally, other greenhouse gases, such as methane, are included by calculating their CO_2 equivalent (eq). There are a lot of theoretical studies that calculate GHG emissions based on the LCA method (e.g. Asdrubali et al. 2015; Carlsson 2014; Goldstein et al. 2011). An analysis of the actual direct emissions of different geothermal power plants in the USA obtains 91 g CO_2 eq/kWh on average (Bloomfield et al. 2003). High emissions are most likely caused by open-loop facilities with high dissolved CO_2 concentration in the brine released due to production (Bertani and Thain 2002). However, Kapetaki et al. (2020) show that binary cycles, for example, EGS Organic Rankine Cycles (ORC), have nearly zero direct emissions. Considering not only electric energy production, but also geothermal heating, Frick et al. (2010) set up a model of a binary power plant and analyze the emissions for different cases and scenarios. Their best-case model has about 7 g CO_2 eq/kWh emissions, whereas the worst-case model has up to 750 g CO_2 eq/kWh. Besides, their model shows that the main GHG emissions result from the construction phase and from drilling. This is also found by Pratiwi et al. (2018), who present an LCA study of different EGS plant scenarios based on the actual plants in Rittershoffen and Illkirch, France. The CO_2 emissions of geothermal power plants are usually lower than the emissions of fossil power plants. In comparison, electricity production from lignite, for example in the Rhenish lignite-mining region (Block F and G in Neurath, Germany) emits 940 g CO_2 eq/kWh (Umweltbundesamt 2017). Power plants using natural gas emit about 500 g CO_2 eq/kWh including the indirect emissions (Amponsah et al. 2014). Space and water heating using oil boilers produces between 310 and 550 g CO_2 eq/kWh, and with gas boilers 210–380 g CO_2 eq/kWh (Parliamentary Office of Science & Technology 2016). In conclusion, energy produced from binary (closed cycle) geothermal sources produces less GHG emissions compared to fossil fuels. However, the emissions from geothermal systems can vary a lot depending on plant type and geological setting. The reservoir brine should be circulated in closed-loop systems to avoid the release of dissolved GHG. The biggest influence on the total GHG emissions of closed-loop systems is the construction phase, and in particular the drilling of wells.

References

Amponsah NY, Troldborg M, Kington B, Aalders I, Hough RL (2014) Greenhouse gas emissions from renewable energy sources: a review of lifecycle considerations. Renew Sustain Energy Rev 39:461–475. https://doi.org/10.1016/j.rser.2014.07.087
Angelino L (2017) Policy and regulation for geothermal energy in the EU. In: Perspectives for geothermal energy in Europe, World Scientific, London
Asdrubali F, Baldinelli G, D'Alessandro F, Scrucca F (2015) Life cycle assessment of electricity production from renewable energies: review and results harmonization. Renew Sustain Energy Rev 42:1113–1122. https://doi.org/10.1016/j.rser.2014.10.082
Bertani R, Thain I (2002) Geothermal power generating plant CO_2 emission survey. International Geothermal Association (IGA) News, pp 1–3
Blanc I, Damen L, Douziech M, Fiaschi D, Harcouet-Menou V, Manfrida G, Mendecka B, Parisi M, Perez Lopez P, Ravier G, Tosti L (2020) First version of harmonized guidelines to perform environmental assessment for geothermal systems based on LCA and non LCA impact indicators: LCA guidelines for geothermal installations. GEOENVI Deliverable D3.2, The GEOENVI Consortium. https://www.geoenvi.eu/wp-content/uploads/2020/07/D3.2_Environmental-impact-and-LCA-Guidelines-for-Geothermal-Installations-v2.pdf. Accessed 8 Oct 2020
Bloomfield K, Moore J, Neilson R (2003) Geothermal energy reduces greenhouse gases. Geotherm Resour Counc Bull 32:77–79

Candela T, van der Veer EF, Fokker PA (2018) On the importance of thermo-elastic stressing in injection-induced earthquakes. Rock Mechan Rock Eng 51(12):3925–3936. https://doi.org/10.1007/s00603-018-1619-6

Carlsson Je (2014) ETRI 2014: energy technology reference indicator projections for 2010-2050. Science and Policy Report, European Commission Joint Research Centre, Institute for Energy and Transport, Luxembourg. https://data.europa.eu/doi/10.2790/057687. Accessed 25 Oct 2021

Corsi R (1986) Scaling and corrosion in geothermal equipment: problems and preventive measures. Geothermics 15(5–6):839–856. https://doi.org/10.1016/0375-6505(86)90097-0

Cuenot N, Frogneux M, Dorbath C, Calò M (2011) Induced microseismic activity during recent circulation tests at the EGS site of Soultz-sous-Forêts (France). In: PROCEEDINGS, Thirty-Sixth workshop on geothermal reservoir engineering, Stanford University, Stanford, California, January 31 - February 2, 2011. https://pangea.stanford.edu/ERE/pdf/IGAstandard/SGW/2011/cuenot.pdf. Accessed 27 Sept 2020

Edwards B, Kraft T, Cauzzi C, Kastli P, Wiemer S (2015) Seismic monitoring and analysis of deep geothermal projects in St Gallen and Basel, Switzerland. Geophys J Int 201(2):1022–1039. https://doi.org/10.1093/gji/ggv059

Ellsworth WL, Giardini D, Townend J, Ge S, Shimamoto T (2019) Triggering of the Pohang, Korea, Earthquake (Mw 5.5) by enhanced geothermal system stimulation. Seismolog Res Lett 90(5):1844–1858. https://doi.org/10.1785/0220190102

European Commission (2020) A renovation wave for Europe – Greening our buildings, creating jobs, improving lives. EUR-Lex 662 final, European Commission. https://eur-lex.europa.eu/legal-content/EN/TXT/?qid=1603122220757&uri=CELEX:52020DC0662. Accessed 24 Sept 2021

Foulger GR, Wilson MP, Gluyas JG, Julian BR, Davies RJ (2018) Global review of human-induced earthquakes. Earth-Sci Rev 178:438–514. https://doi.org/10.1016/j.earscirev.2017.07.008

Frick S, Kaltschmitt M, Schröder G (2010) Life cycle assessment of geothermal binary power plants using enhanced low-temperature reservoirs. Energy 35(5):2281–2294. https://doi.org/10.1016/j.energy.2010.02.016

Garcia AV, Thomsen K, Stenby EH (2005) Prediction of mineral scale formation in geothermal and oilfield operations using the extended UNIQUAC model: part I. Sulfate scaling minerals. Geothermics 34(1):61–97. https://doi.org/10.1016/j.geothermics.2004.11.002

Gehringer M, Loksha V (2012) Geothermal handbook: planning and financing power generation. Technical report 002/12, World Bank, Washington, DC. https://openknowledge.worldbank.org/handle/10986/23712. Accessed 19 Aug 2020

Goldstein B, Hiriart G, Bertani R, Bromley C, Guitérrez-Negrin L, Huenges F, Muraoka H, Ragnarsson A, Tester J, Zui V (2011) Geothermal energy. In: IPCC special report on renewable energy sources and climate change mitigation, Cambridge University Press, Cambridge, UK. http://www.ipcc-wg3.de/srren-report/

Grigoli F, Cesca S, Rinaldi AP, Manconi A, López-Comino JA, Clinton JF, Westaway R, Cauzzi C, Dahm T, Wiemer S (2018) The November 2017 Mw 5.5 Pohang earthquake: a possible case of induced seismicity in South Korea. Science 360(6392):1003–1006. https://doi.org/10.1126/science.aat2010

Le Guénan T, Hamm V, Calcagno P, Loschetter A (2020) Risk register. GEORISK deliverable D2.1, The GEORISK consortium. https://www.georisk-project.eu/wp-content/uploads/2019/04/GEORISK-D-2-1-risk-register.pdf. Accessed 21 Oct 2021

Hillers G, Vuorinen TAT, Uski MR, Kortström JT, Mäntyniemi PB, Tiira T, Malin PE, Saarno T (2020) The 2018 geothermal reservoir stimulation in Espoo/Helsinki, Southern Finland: seismic network anatomy and data features. Seismolog Res Lett 91(2A):770–786. https://doi.org/10.1785/0220190253

Hofmann H, Zimmermann G, Farkas M, Huenges E, Zang A, Leonhardt M, Kwiatek G, Martinez-Garzon P, Bohnhoff M, Min KB, Fokker P, Westaway R, Bethmann F, Meier P, Yoon KS, Choi JW, Lee TJ, Kim KY (2019) First field application of cyclic soft stimulation at the Pohang enhanced geothermal system site in Korea. Geophys J Int 217(2):926–949. https://doi.org/10.1093/gji/ggz058

International Organization for Standardization (2018) ISO 31000: Risk management: principles and guidelines. International Organization for Standardization, Geneva, Switzerland

IRENA (2021) World energy transition outlook: 1.5 °C pathway. International Renewable Energy Agency (IRENA), Abu Dhabi. https://www.irena.org/publications/2021/Jun/World-Energy-Transitions-Outlook. Accessed 23 Oct 2021

Kapetaki Z, Ruiz Castello P, Armani R, Bodis K, Fahl F, Gonzalez Aparicio I, Jaeger-Waldau A, Lebedeva N, Pinedo Pascua I, Scarlat N, Taylor N, Telsnig T, Uihlein A, Vazquez Hernandez C, Zangheri P (2020) Clean energy technologies in coal regions. EUR - Scientific and Technical Research Reports, Publications Office of the European Union. https://doi.org/10.2760/063496 (online), https://doi.org/10.2760/384605 (print)

Kim KH, Seo W, Han J, Kwon J, Kang SY, Ree JH, Kim S, Liu K (2020) The 2017 ML 5.4 Pohang earthquake sequence, Korea, recorded by a dense seismic network. Tectonophysics 774:228306. https://doi.org/10.1016/j.tecto.2019.228306

Kukkonen IT, Pentti M (2021) St1 Deep Heat Project: Geothermal energy to the district heating network in Espoo. IOP Conf Ser: Earth Environ Sci 703(1):012035. https://doi.org/10.1088/1755-1315/703/1/012035

Leonhardt M, Kwiatek G, Martínez-Garzón P, Bohnhoff M, Saarno T, Heikkinen P, Dresen G (2021) Seismicity during and after stimulation of a 6.1 km deep enhanced geothermal system in Helsinki, Finland. Solid Earth 12(3):581–594. https://doi.org/10.5194/se-12-581-2021

Leucht M, Kölbel T, Laborgne P, Khomenko N (2010) The role of societal acceptance in renewable energy innovations' breakthrough in the case of deep geothermal technology. In: Proceedings World geothermal congress 2010, Bali, Indonesia, p 12. https://www.geothermal-energy.org/pdf/IGAstandard/WGC/2010/0219.pdf. Accessed 24 Oct 2021

Manzella A, Allansdottir A, Pellizzone A (2019) Geothermal energy and society. Lecture notes in energy, vol 67. Springer International Publishing, Cham. https://doi.org/10.1007/978-3-319-78286-7

Mignan A, Broccardo M, Wiemer S, Giardini D (2017) Induced seismicity closed-form traffic light system for actuarial decision-making during deep fluid injections. Sci Rep 7(1):13607. https://doi.org/10.1038/s41598-017-13585-9

Mukuhira Y, Asanuma H, Niitsuma H, Häring MO (2013) Characteristics of large-magnitude microseismic events recorded during and after stimulation of a geothermal reservoir at Basel, Switzerland. Geothermics 45:1–17. https://doi.org/10.1016/j.geothermics.2012.07.005

Parliamentary Office of Science & Technology (2016) Carbon footprint of heat generation. Post Note 523, Houses of Parliament, Parliamentary Office of Science & Technology. https://researchbriefings.files.parliament.uk/documents/POST-PN-0523/POST-PN-0523.pdf. Accessed 26 Oct 2021

Pellizzone A, Allansdottir A, Muttoni G, Franco RD, Manzella A (2015) Social Acceptance of Geothermal Energy in Southern Italy. Proceedings world geothermal congress 2015, Melbourne, Australia, 19-25 April 2015 p 6

Pluymakers A (2019) Geothermal energy and structural geology? https://blogs.egu.eu/divisions/ts/2019/10/25/geothermal-energy-and-structural-geology/. Accessed 25 Oct 2021

Porter RTJ, Striolo A, Mahgerefteh H, Walker JF (2019) Addressing the risks of induced seismicity in subsurface energy operations. WIREs Energy Environ 8(2):e324. https://doi.org/10.1002/wene.324

Pratiwi A, Ravier G, Genter A (2018) Life-cycle climate-change impact assessment of enhanced geothermal system plants in the Upper Rhine Valley. Geothermics 75:26–39. https://doi.org/10.1016/j.geothermics.2018.03.012

Reith S, Kölbel T, Schlagermann P, Pellizzone A, Allansdottir A (2013) Public acceptance of geothermal electricity production. GEOELEC Deliverable Deliverable 4.4, The GEO-ELEC Consortium. http://www.geoelec.eu/wp-content/uploads/2014/03/D-4.4-GEOELEC-report-on-public-acceptance.pdf. Accessed 26 Oct 2021

Rosberg JE, Erlström M (2019) Evaluation of the Lund deep geothermal exploration project in the Romeleåsen Fault Zone, South Sweden: a case study. Geotherm Energy 7(1). https://doi.org/10.1186/s40517-019-0126-7

Sanyal SK, Robertson-Tait A, Jayawardena MS, Huttrer G, Berman L (2016) Comparative analysis of approaches to geothermal ressource risk mitigation: a global survey. Technical report 024/16, The International Bank for Reconstruction and Development/ The World Bank Group. http://documents1.worldbank.org/curated/en/621131468180534369/pdf/105172-ESM-P144569-PUBLIC-FINAL-ESMAP-GeoRiskMitigation-KS024-16-web.pdf. Accessed 25 Oct 2021

Schmittbuhl J, Lambotte S, Lengliné O, Grunberg M, Jund H, Vergne J, Cornet F, Doubre C, Masson F (2021) Induced and triggered seismicity below the city of Strasbourg, France from November 2019 to January 2021. Comptes Rendus Géoscience Online first. https://doi.org/10.5802/crgeos.71

Sperber A, Moeck I, Brandt W (2010) Drilling into Geothermal reservoirs (chap 3). In: Geothermal energy systems. Wiley, New York, pp 113–171. https://doi.org/10.1002/9783527630479.ch3

Umweltbundesamt (2017) Daten und Fakten zu Braun- und Steinkohlen. Technical report Dezember 2017, Umweltbundesamt, https://www.umweltbundesamt.de/sites/default/files/medien/1410/publikationen/171207_uba_hg_braunsteinkohle_bf.pdf. Accessed 27 Oct 2021

Wilson M, Foulger G, Gluyas J, Davies R, Julian B (2017) Hiquake: the human-induced earthquake database. Seismolog Res Lett 88(6):1560–1565

Xu T, Pruess K (2001) Modeling multiphase non-isothermal fluid flow and reactive geochemical transport in variably saturated fractured rocks: 1. Methodology. Amer J Sci 301(1):16–33. https://doi.org/10.2475/ajs.301.1.16

Zang A, Oye V, Jousset P, Deichmann N, Gritto R, McGarr A, Majer E, Bruhn D (2014) Analysis of induced seismicity in geothermal reservoirs – an overview. Geothermics 52:6–21. https://doi.org/10.1016/j.geothermics.2014.06.005

Zarrouk SJ, McLean K (2019) Geothermal well test analysis - Fundamentals, applications and advanced techniques. Elsevier, Amsterdam

Chapter 5
Summary and Conclusions

In the context of combating climate change, there are currently many initiatives to decarbonize the heating sector, both at national and European level. Geothermal energy is a clean and sustainable source of heat that can be extracted from the sub-surface by **drilling** (Sect. 3.3.1) and by pumping hot fluid (usually water or steam) to the surface. The hot fluid can either be used for heating purposes (e.g. for build-ings and industrial processes) or for electricity generation. Here, we focus on fluid temperatures which qualify for direct heating.

Unfortunately, drilling into the Earth's crust is very expensive. For deep geother-mal wells, drilling often comprises more than 50% of the total cost (Garabetian 2019). As a result, the commercial use of geothermal energy concentrates not just in Europe on areas with high **geothermal potential** (Sect. 2.1), that is, areas with favorable conditions such as high temperatures in shallow depths. Besides the very favorable magmatic and volcanic areas (e.g. in Italy and Iceland), geothermal energy in Europe is mainly used from **hydrothermal reservoirs** (Sect. 1.5.2) in several basin settings across Europe (e.g. the Paris, Molasse, and Pannonian Basins). The wells drilled into these **conventional geothermal systems** for commercial use of geothermal energy are rarely deeper than 3 km.

As these conventional systems are geographically and quantitatively limited, the use of geothermal energy in regions with medium or low geothermal potential is becoming interesting for a sustainable energy supply and especially for the decar-bonization of the heating sector. The development of these resources is essential for unleashing the full potential of geothermal energy, since they make up more than 85% of the total geothermal potential in Central Europe and other countries (Richter 2017). To achieve the necessary temperatures in such less favorable regions, wells have to be drilled into deep basement rocks (usually deeper than 3 km). As the per-meability of these basement rocks is too low for economic use, their geothermal exploitation requires the enhancement of the reservoir, i.e. the artificial **stimulation** (Sect. 3.4.1) for increasing permeability and accessing the heat stored in the rocks.

© The Author(s), under exclusive license to Springer Nature Switzerland AG 2022
J. Fink et al., *State of the Art in Deep Geothermal Energy in Europe*,
SpringerBriefs in Earth System Sciences,
https://doi.org/10.1007/978-3-030-96870-0_5

Exploiting such **petrothermal or enhanced geothermal systems** (EGS, Sect. 1.5.2) is still considered unconventional or innovative technology. Although some EGS projects exist in Europe, it is not state of the art or commercial practice, they are mostly research and pilot projects. Moreover, existing EGS projects are limited to regions with a relatively high geothermal potential (high geothermal gradient) such as the Upper Rhine Graben. Only recently, first projects like the Okaniemi project in Finland (Kukkonen and Pentti 2021) started to develop deep geothermal resources using the EGS technology in areas with low geothermal gradients. Both ongoing and abandoned EGS projects in Europe are listed in Appendix 2.

. The EGS technology can also be applied to enhance **hot sedimentary aquifers** (in- or near-field EGS) but the abundance of hot sedimentary aquifers is much smaller than that of petrothermal (EGS) systems (green-field EGS). However, the abundance inversely correlates with the associated risk of failure. Therefore, it is paramount to develop and apply technologies to increase the probability of success of geothermal wells tapping into deep petrothermal systems. So far, creating petrothermal reservoirs with sufficient permeability to allow for commercial flow rates has proven difficult. Without a successfully stimulated reservoir, a 5 km deep doublet has limited commercial value (Sigfússon and Uihlein 2015). Therefore, improved understanding of the thermo-hydro-mechanical and chemical (THM/C) processes and development of effective and safe stimulation methods is a fundamental issue for EGS. Indeed, THM/C modeling can be a key tool for predicting the lifetime of a geothermal reservoir and for assessing risks associated with exploitation.

Hydraulic stimulation (Sect. 3.4.1) remains the most important technique for enhancing a geothermal reservoir, i.e. for increasing its permeability by creating new or stimulating existing fractures. This process induces seismicity, which constitutes the largest risk factor associated with EGS while at the same time being the tool for creating the geothermal reservoir. The EGS project in Basel (Switzerland) was abandoned because micro- to minor-seismic events occurred during hydraulic stimulation that led to public debate and a negative public perception of the project. Therefore, with regard to **induced seismicity**

1. risk assessment should be included into resource assessment,
2. both reservoir stimulation and operation should be monitored using state-of-the-art technology,
3. innovative stimulation techniques such as soft cyclic stimulation (Hofmann et al. 2018, 2019) should be applied and further developed, and
4. public relations work regarding the opportunities and risks of geothermal energy should accompany the project from the very beginning.

Besides hydraulic stimulation, thermal and chemical stimulation can be applied to improve the geothermal circulation. However, they involve risks, too. For cost-effective development of EGS resources as well as for the reduction of the involved **risks** (Chap. 4), all phases of the geothermal project require innovation. We introduce conventional as well as innovative technologies for exploration, assessment, engineering, and development of deep geothermal resources in Chap. 3.

To conclude, we would like to emphasize that using energy from deep (>3 km) geothermal projects in regions with medium or low geothermal potential that require reservoir enhancement, is not yet a market-ready technology but it has significant potential for decarbonizing the heating sector. Currently, "In the case of district heating, geographical alignment of resources with market demand centers is a key limiting factor for development" of geothermal energy. In prospect, "Enabling cost-effective development of EGS resources through technology improvements can reduce geographic limitations on geothermal district heating". Both statements of the (U.S. Department of Energy 2019) apply to Europe, too. Innovation is required primarily for reducing costs and risks, especially for deep drilling in hard rock and risks associated with reservoir stimulation.

Further development of the EGS technology requires demonstration and pilot projects that involve both the drilling to reservoir depth and the stimulation of the reservoir. So far, the geothermal sector in Europe lacks successful deep EGS projects. A proof of the EGS concept in a deep hard-rock formation is needed for convincing investors and decision-makers, and for increasing public acceptance.

References

Garabetian T (2019) Report on competitiveness of the geothermal industry. Technical report, European technology and innovation platform on deep geothermal (ETIP-DG). http://www.etip-dg. eu/front/wp-content/uploads/D4.6-Report-on-Competitiveness.pdf. Accessed 25 Oct 2021

Hofmann H, Zimmermann G, Zang A, Min KB (2018) Cyclic soft stimulation (CSS): a new fluid injection protocol and traffic light system to mitigate seismic risks of hydraulic stimulation treatments. Geotherm Energy 6(1):27. https://doi.org/10.1186/s40517-018-0114-3

Hofmann H, Zimmermann G, Farkas M, Huenges E, Zang A, Leonhardt M, Kwiatek G, Martinez-Garzon P, Bohnhoff M, Min KB, Fokker P, Westaway R, Bethmann F, Meier P, Yoon KS, Choi JW, Lee TJ, Kim KY (2019) First field application of cyclic soft stimulation at the Pohang enhanced geothermal system site in Korea. Geophys J Int 217(2):926–949. https://doi.org/10. 1093/gji/ggz058

Kukkonen IT, Pentti M (2021) St1 Deep Heat Project: Geothermal energy to the district heating network in Espoo. IOP Conf Ser: Earth Environ Sci 703(1):012035. https://doi.org/10.1088/ 1755-1315/703/1/012035

Richter M (2017) Summary of new drilling technologies. Technical report, International energy agency - geothermal (IEA Geothermal). http://iea-gia.org/wp-content/uploads/2014/10/IEA-Geothermal-Drilling-Technologies.pdf. Accessed 25 Oct 2021

Sigfússon B, Uihlein A (2015) 2015 JRC geothermal energy status report - technology, market and economic aspects of geothermal energy in Europe. Technical report, Institute for Energy and Transport (Joint Research Centre), Publications Office of the European Union. https://doi.org/ 10.2790/959587

US Department of Energy (2019) GeoVision - Harnessing the heat beneath our feet. Technical report, U.S. Department of Energy. http://www.osti.gov/scitech. Accessed 22 Oct 2021

Index

Printed in the United States
by Baker & Taylor Publisher Services